W9-BYB-823

LOGIC
MADE
EASY

Also by Deborah J. Bennett

Randomness

LOGIC MADE EASY

How to Know When Language Deceives You

DEBORAH J. BENNETT

W · W · NORTON & COMPANY  NEW YORK LONDON

Printed in the United States of America
First Edition

Manufacturing by The Haddon Craftsmen, Inc.
Book design by Margaret M. Wagner
Production manager: Julia Druskin

Library of Congress Cataloging-in-Publication Data

Bennett, Deborah J., 1950–
Logic made easy : how to know when language deceives you /
Deborah J. Bennett.— 1st ed.
p. cm.
Includes bibliographical references and index.
ISBN 0-393-05748-8
1. Reasoning. 2. Language and logic. I. Title.
BC177 .B42 2004
160—dc22

2003026910

W. W. Norton & Company, Inc., 500 Fifth Avenue, New York, N.Y. 10110
www.wwnorton.com

W. W. Norton & Company Ltd., Castle House, 75/76 Wells Street, London W1T 3QT

1 2 3 4 5 6 7 8 9 0

Contents

Introduction: Logic Is Rare

Crime is common. Logic is rare.
Sherlock Holmes
in *The Adventure of the Copper Beeches*

Logic Made Easy is a book for anyone who believes that logic is rare. It is a book for those who think they are logical and wonder why others aren't. It is a book for anyone who is curious about why logical thinking doesn't come "naturally." It is a book for anyone who wants to be more logical. There are many fine books on the rules of logic and the history of logic, but here you will read the story of the barriers we face in trying to communicate logically with one another.

It may surprise you to learn that logical reasoning is difficult. How can this be? Aren't we all logical by virtue of being human? Humans are, after all, *reasoning* animals, perhaps the only animals capable of reason. From the time we are young children, we ask Why?, and if the answer doesn't make sense we are rarely satisfied. What does "make sense" mean anyway? Isn't "makes sense" another way of saying "is logical"?

Children hold great stock in rules being applied fairly and rules that make sense. Adults, as well, hold each other to the standards of consistency required by logic. This book is for anyone who thinks being logical is important. It is also for anyone who needs to be convinced that logic is important.

To be considered illogical or inconsistent in our positions or behaviors is insulting to us. Most of us think of ourselves as being logical. Yet the evidence indicates something very different. It turns out that we are often not very logical. Believing ourselves to be logical is common, but logic itself is rare.

This book is unlike other books on logic. Here you will learn why logical reasoning isn't so easy after all. If you think you are fairly logical, try some of the logic puzzles that others find tricky. Even if you don't fall into the trap of faulty reasoning yourself, this book will help you understand the ways in which others encounter trouble.

If you are afraid that you are not as logical as you'd like to be, this book will help you see why that is. Hopefully, after reading this book you will be more logical, more aware of your language. There is an excellent chance that your thinking will be clearer and your ability to make your ideas clearer will be vastly improved. Perhaps most important, you will improve your capability to evaluate the thinking and arguments of others—a tool that is invaluable in almost any walk of life.

We hear logical arguments every day, when colleagues or friends try to justify their thoughts or behaviors. On television, we listen to talking heads and government policy-makers argue to promote their positions. Virtually anyone who is listening to another argue a point must be able to assess what assumptions are made, follow the logic of the argument, and judge whether the argument and its conclusion are valid or fallacious.

Assimilating information and making inferences is a basic component of the human thought process. We routinely make logical inferences in the course of ordinary conversation, reading, and listening. The concept that certain statements necessarily do or do not follow from certain other statements is at the core of our reasoning abilities. Yet, the rules of language and logic oftentimes seem at odds with our intuition.

Many of the mistakes we make are caused by the ways we use language. Certain nuances of language and semantics get in the way of "correct thinking." This book is not an attempt to delve deeply into the study of semantics or cognitive psychology. There are other comprehensive scholarly works in those fields. *Logic Made Easy* is a down-to-earth story of logic and language and how and why we make mistakes in logic.

In Chapter 2, you will discover that philosophers borrowed from ideas of mathematical proof as they became concerned about mistakes in logic in their never-ending search for truth. In Chapters 3, 4, and 5, as we begin to explore the language and vocabulary of logical statements—simple vocabulary like *all*, *not*, and *some*—you will find out (amazingly enough) that knowledge, familiarity, and truth can interfere with logic. But how can it be easier to be logical about material you know nothing about?

Interwoven throughout the chapters of this book, we will learn what history has to offer by way of explanation of our difficulties in reasoning logically. Although rules for evaluating valid arguments have been around for over two thousand years, the common logical fallacies identified way back then remain all too common to this day. Seemingly simple statements continue to trip most people up.

The Mistakes We Make

While filling out important legal papers and income tax forms, individuals are required to comprehend and adhere to formally written exacting language—and to digest and understand the fine print, at least a little bit. Getting ready to face your income tax forms, you encounter the statement "All those who reside in New Jersey must fill out Form 203." You do not live in New Jersey. Do you have to fill out Form 203? Many individuals who consider themselves logical might answer no to this question. The correct answer is "We don't know—maybe, maybe not. There is not enough information." If the statement had read "Only those who reside in New Jersey must fill out Form 203" and you aren't a New Jersey resident, then you would be correct in answering no.

Suppose the instructions had read "Only those who reside in New Jersey should fill out Form 203" and you are from New Jersey. Do you have to fill out Form 203? Again, the correct answer is "Not enough information. Maybe, maybe not." While only New Jersey residents need to fill out the form, it is not necessarily true that all New Jersey-ites must complete it.

Our interpretations of language are often inconsistent. The traffic information sign on the expressway reads "Delays until exit 26." My husband seems to speed up, saying that he can't wait to see if they are lying. When I inquire, he says that there should be no delays after exit 26. In other words, he interprets the sign to say "Delays until exit 26 and no delays thereafter." On another day, traffic is better. This time the sign reads "Traffic moving well to exit 26." When I ask him what he thinks will happen after exit 26, he says that there may be traffic or there may not. He believes the sign information is only current up to exit 26. Why does he interpret the language on the sign as a promise about what will happen beyond exit 26 on the one hand, and no promise at all on the other?

Cognitive psychologists and teachers of logic have often observed that mistakes in inference and reasoning are not only extremely common but also nearly always of a particular kind.

Most of us make mistakes in reasoning; we make similar mistakes; and we make them over and over again.

Beginning in the 1960s and continuing to this day, there began an explosion of research by cognitive psychologists trying to pin down exactly why these mistakes in reasoning occur so often. Experts in this area have their own journals and their own professional societies. Some of the work in this field is revealing and bears directly on when and why we make certain errors in logic.

Various logical "tasks" have been devised by psychologists trying to understand the reasoning process and the source of our errors in reasoning. Researchers Peter C. Wason and Philip Johnson-Laird claim that one particular experiment has an almost hypnotic effect on some people who try it, adding that this experiment tempts the majority of subjects into an interesting and deceptively fallacious inference. The subject is shown four colored symbols: a blue diamond, a yellow diamond, a blue circle, and a yellow circle. (See Figure 1.) In one version of the problem, the experimenter gives the following instructions:

> I am thinking of one of those colors and one of those shapes. If a symbol has either the color I am thinking about, or the shape I am thinking about, or both, then I *accept* it, but otherwise I *reject* it. I *accept* the blue diamond. Does anything follow about my acceptance, or rejection, of the other symbols?[1]

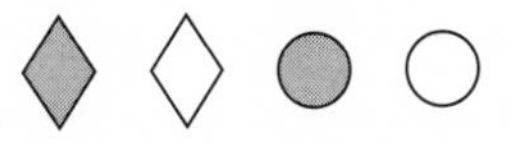

Figure 1. "Blue diamond" experiment.

A mistaken inference characteristically made is to conclude that the yellow circle will be rejected. However, that can't be right. The blue diamond would be accepted if the experimenter were thinking of "blue and circle," in which case the yellow circle would not be rejected. In accepting the blue diamond, the experimenter has told us that he is thinking of (1) blue and diamond, (2) blue and circle, or (3) yellow and diamond, but we don't know which. Since he accepts all other symbols that have either the color *or* the shape he is thinking about (and otherwise rejects the symbol), in case 1 he accepts all blue shapes and any color diamond. (He rejects only the yellow circle.) In case 2, he accepts all blue shapes and any color circle. (He rejects only the yellow diamond.) In case 3, he accepts any yellow shapes and any color diamonds. (He rejects only the blue circle.) Since we don't know which of the above three scenarios he is thinking of, we can't possibly know which of the other symbols will be rejected. (We do know, however, that *one* of them will be.) His acceptance of the blue diamond does not provide enough information for us to be certain about his acceptance or rejection of any of the other symbols. All we know is that two of the others will be accepted and one will be rejected. The only inference that we can make concerns what the experimenter is thinking—or rather, what he is not thinking. He is not thinking "yellow and circle."[2]

As a college professor, I often witness mistakes in logic. Frequently, I know exactly which questions as well as which wrong answers will tempt students into making errors in logical thinking. Like most teachers, I wonder, Is it me? Is it only my students? The answer is that it is not at all out of the ordinary to find even intelligent adults making mistakes in simple deductions.

Several national examinations, such as the Praxis I™ (an examination for teaching professionals), the Graduate Records Examination (GRE®) test, the Graduate Management Admissions Test

(GMAT®), and the Law School Admissions Test (LSAT®), include logical reasoning or analytical questions. It is these types of questions that the examinees find the most difficult.

A question from the national teachers' examination, given in 1992 by the Educational Testing Service (ETS®), is shown in Figure 2.[3] Of the 25 questions on the mathematics portion of this examination, this question had the lowest percentage of correct responses. Only 11 percent of over 7,000 examinees could answer the question correctly, while the vast majority of the math questions had correct responses ranging from 32 percent to 89 percent.[4] Ambiguity may be the source of some error here. The first two given statements mention education *majors* and the third given statement switches to a statement about mathematics *students*. But, most probably, those erring on this question were

---

Given:

1. All education majors student teach.
2. Some education majors have double majors.
3. Some mathematics students are education majors.

Which of the following conclusions necessarily follows from 1, 2, and 3 above?

A. Some mathematics students have double majors.
B. Some of those with double majors student teach.
C. All student teachers are education majors.
D. All of those with double majors student teach.
E. Not all mathematics students are education majors.

Figure 2. A sample test question from the national teachers' examination, 1992. (*Source:* The Praxis Series: Professional Assessments for Beginning Teachers® NTE Core Battery Tests Practice and Review [1992]. Reprinted by permission of Educational Testing Service, the copyright owner.)

seduced by the truth of conclusion C. It may be a true conclusion, but it does not necessarily follow from the given statements. The correct answer, B, logically follows from the first two given statements. Since *all* education majors student teach and some of that group of education majors have double majors, it follows that some with double majors student teach.

For the past twenty-five years, the Graduate Records Examination (GRE) test given by the Educational Testing Service (ETS) consisted of three measures—verbal, quantitative, and analytical. The ETS indicated that the analytical measure tests our ability to understand relationships, deduce information from relationships, analyze and evaluate arguments, identify hypotheses, and draw sound inferences. The ETS stated, "Questions in the analytical section measure reasoning skills developed in virtually all fields of study."[5]

Logical and analytical sections comprise about half of the LSAT, the examination administered to prospective law school students. Examinees are expected to analyze arguments for hidden assumptions, fallacious reasoning, and appropriate conclusions. Yet, many prospective law students find this section to be extremely difficult.

Logic Should Be Everywhere

It is hard to imagine that inferences and deductions made in daily activity aren't based on logical reasoning. A doctor must reason from the symptoms at hand, as must the car mechanic. Police detectives and forensic specialists must process clues logically and reason from them. Computer users must be familiar with the logical rules that machines are designed to follow. Business decisions are based on a logical analysis of actualities and contingencies. A juror must be able to weigh evidence and follow the logic of an attorney prosecuting or defending a case: If the defendant was at the movies at the time, then he couldn't have committed the crime. As a matter of fact, *any* problem-solving activity, or what educators today call *critical thinking*, involves pattern-seeking and conclusions arrived at through a logical path.

Deductive thinking is vitally important in the sciences, with the rules of inference integral to forming and testing hypotheses. Whether performed by a human being or a computer, the procedures of logical steps, following one from another, assure that the conclusions follow validly from the data. The certainty that logic provides makes a major contribution to our discovery of truth. The great mathematician, Leonhard Euler (pronounced *oiler*) said that logic "is the foundation of the certainty of all the knowledge we acquire."[6]

Much of the history of the development of logic can shed light on why many of us make mistakes in reasoning. Examining the roots and evolution of logic helps us to understand why so many of us get tripped up so often by seemingly simple logical deductions.

How History Can Help

Douglas Hofstadter, author of *Gödel, Escher, and Bach*, said that the study of logic began as an attempt to mechanize the thought processes of reasoning. Hofstadter pointed out that even the ancient Greeks knew "that reasoning is a patterned process, and is at least partially governed by statable laws."[7] Indeed, the Greeks believed that deductive thought had patterns and quite possibly laws that could be articulated.

Although certain types of discourse such as poetry and storytelling may not lend themselves to logical inquiry, discourse that requires *proof* is fertile ground for logical investigation. To prove a statement is to infer the statement validly from known or accepted truths, called *premises*. It is generally acknowledged that the earliest application of proof was demonstrated by the Greeks in mathematics—in particular, within the realm of geometry.

While a system of formal deduction was being developed in geometry, philosophers began to try to apply similar rules to metaphysical argument. As the earliest figure associated with the logical argument, Plato was troubled by the arguments of the Sophists. The Sophists used deliberate confusion and verbal tricks in the course of a debate to win an argument. If you were un*sophist*icated, you might be fooled by their arguments.[8] Aristotle, who is considered the inventor of logic, did not resort to the language tricks and ruses of the Sophists but, rather, attempted to systematically lay out rules that all might agree dealt exclusively with the correct usage of certain statements, called *propositions*.

The vocabulary we use within the realm of logic is derived directly from Latin translations of the vocabulary that Aristotle used when he set down the rules of logical deduction through propositions. Many of these words have crept into our everyday language. Words such as *universal* and *particular*, *premise* and *conclusion*, *contradictory* and *contrary* are but a few of the terms first introduced by Aristotle that have entered into the vocabulary of all educated persons.

Aristotle demonstrated how sentences could be joined together properly to form valid arguments. We examine these in Chapter 5. Other Greek schools, mainly the Stoics, also contributed a system of logic and argument, which we discuss in Chapters 6 and 7.

At one time, logic was considered one of the "seven liberal arts," along with grammar, rhetoric, music, arithmetic, geometry, and astronomy. Commentators have pointed out that these subjects represented a course of learning deemed vital in the "proper preparation for the life of the ideal knight and as a necessary step to winning a fair lady of higher degree than the suitor."[9] A sixteenth-century logician, Thomas Wilson, includes this verse in his book on logic, *Rule of Reason*, the first known English-language book on logic:

> Grammar doth teach to utter words.
> To speak both apt and plain,
> Logic by art sets forth the truth,
> And doth tell us what is vain.
>
> Rhetoric at large paints well the cause,
> And makes that seem right gay,
> Which Logic spake but at a word,
> And taught as by the way.
>
> Music with tunes, delights the ear,
> And makes us think it heaven,
> Arithmetic by number can make
> Reckonings to be even.
>
> Geometry things thick and broad,
> Measures by Line and Square,
> Astronomy by stars doth tell,
> Of foul and else of fair.[10]

Almost two thousand years after Aristotle's formulation of the rules of logic, Gottfried Leibniz dreamed that logic could become a universal language whereby controversies could be settled in the same exacting way that an ordinary algebra problem is worked out. In Chapter 9 you will find that alone among seventeenth-century philosophers and mathematicians, Leibniz (the co-inventor with Isaac Newton of what we today call *calculus*) had a vision of being able to create a universal language of logic and reasoning from which all truths and knowledge could be derived. By reducing logic to a symbolic system, he hoped that errors in thought could be detected as computational errors. Leibniz conceived of his system as a means of resolving conflicts among peoples—a tool for world peace. The world took little notice of Leibniz's vision until George Boole took up the project some two hundred years later.

Bertrand Russell said that pure mathematics was discovered by George Boole, and historian E. T. Bell maintained that Boole was one of the most original mathematicians that England has produced.[11] Born to the tradesman class of British society, George Boole knew from an early age that class-conscious snobbery would make it practically impossible for him to rise above his lowly shopkeeper station. Encouraged by his family, he taught himself Latin, Greek, and eventually moved on to the most advanced mathematics of his day. Even after he achieved some reputation in mathematics, he continued to support his parents by teaching elementary school until age 35 when Boole was appointed Professor of Mathematics at Queen's College in Cork, Ireland.

Seven years later in 1854, Boole produced his most famous work, a book on logic entitled *An Investigation of the Laws of Thought*. Many authors have noted that "the laws of thought" is an extreme exaggeration—perhaps thought involves more than logic. However, the title reflects the spirit of his intention to give logic the rigor and inevitability of laws such as those that algebra enjoyed.[12] Boole's work is the origin of what is called *Boolean logic*, a system so simple that even a machine can employ its rules. Indeed, today in the age of the computer, many do. You will see in Chapter 10 how logicians attempted to create reasoning machines.

Among the nineteenth-century popularizers of Boole's work in symbolic logic was Rev. Charles Lutwidge Dodgson, who wrote under the pseudonym of Lewis Carroll. He was fascinated by Boole's mechanized reasoning methods of symbolic logic and wrote logic puzzles that could be solved by those very methods. Carroll wrote a two-volume work called *Symbolic Logic* (only the first volume appeared in his lifetime) and dedicated it to the memory of Aristotle. It is said that Lewis Carroll, the author of *Alice's Adventures in Wonderland*, considered his book on logic the work of which he was most proud. In the Introduction of *Symbolic Logic*, Carroll describes, in glowing terms, what he sees as the benefits of studying the subject of logic.

> Once master the machinery of Symbolic Logic, and you have a mental occupation always at hand, of absorbing interest, and one that will be of real *use* to you in *any* subject you take up. It will give you clearness of thought—the ability to *see your way* through a puzzle—the habit of arranging your ideas in an orderly and get-at-able form—and, more valuable than all, the power to detect *fallacies*, and to tear to pieces the flimsy illogical arguments, which you will so continually encounter in books, in newspapers, in speeches, and even in sermons, and which so easily delude those who have never taken the trouble to master this fascinating Art. *Try it*. That is all I ask of you![13]

Carroll was clearly intrigued with Boole's symbolic logic and the facility it brought to bear in solving problems, structuring thoughts, and preventing the traps of illogic.

▸ ▸

The language of logic employs simple everyday words—words that we use all the time and presumably understand. The rules for combining these terms into statements that lead to valid inferences have been around for thousands of years. Are the rules of logic themselves logical? Why do we need rules? Isn't our ability to reason what makes us *human* animals?

Even though we use logic all the time, it appears that we aren't very logical. Researchers have proposed various reasons as to the cause of error in deductive thinking. Some have suggested that individuals ignore available information, add information of their own, have trouble keeping track of information, or are unable to retrieve necessary information.[14] Some have suggested that ordinary language differs from the language used by logicians, but others hypothesize that errors are due to our cognitive inability. Some have suggested that familiarity with the content of an argument enhances our ability to infer correctly, while others have suggested that it is familiarity that interferes with that ability.[15] If the problem is not faulty reasoning, then what is it in the material that causes us to focus our attention on the wrong things?

As we progress through the following chapters, we will examine the ways that we use (or misuse) language and logic in everyday life. What insight can we gain from examining the roots and evolution of logic? How can the psychologists enlighten us about the reasoning mistakes we commonly make? What can we do to avoid the pitfalls of illogic? Can understanding the rules of logic foster clear thinking? Perhaps at the journey's end, we will all be thinking more logically.

But let's not get ahead of ourselves; let us start at the beginning. What is the minimum we expect from each other in terms of logical thinking? To answer that question, we need to examine the roots of logic that are to be found in the very first glimmerings of mathematical proof.

LOGIC MADE EASY

1

PROOF

No amount of experimentation can ever prove me right;
a single experiment can prove me wrong.
ALBERT EINSTEIN

Consistency Is All I Ask

There are certain principles of ordinary conversation that we expect ourselves and others to follow. These principles underlie all reasoning that occurs in the normal course of the day and we expect that if a person is honest and reasonable, these principles will be followed. The guiding principle of rational behavior is consistency. If you are consistently consistent, I trust that you are not trying to pull the wool over my eyes or slip one by me.

If yesterday you told me that you loved broccoli and today you claim to hate it, because I know you to be rational and honest I will probably conclude that something has changed. If nothing has changed then you are holding inconsistent, contradictory positions. If you claim that you always look both ways before crossing the street and I see you one day carelessly ignoring the traffic as you cross, your behavior is contradicting your claim and you are being inconsistent.

These principles of consistency and noncontradiction were

recognized very early on to be at the core of mathematical proof. In *The Topics*, one of his treatises on logical argument, Aristotle expresses his desire to set forth methods whereby we shall be able "to reason from generally accepted opinions about any problem set before us and shall ourselves, when sustaining an argument, avoid saying anything self-contradictory."[1] To that end, let's consider both the *law of the excluded middle* and the *law of noncontradiction*—logical truisms and the most fundamental of axioms. Aristotle seems to accept them as general principles.

The law of the excluded middle requires that a thing must either possess a given attribute or must not possess it. A thing must be one way or the other; there is no middle. In other words, the middle ground is excluded. A shape either is a circle or is not a circle. A figure either is a square or is not a square. Two lines in a plane either intersect or do not intersect. A statement is either true or not true. However, we frequently see this principle misused.

How many times have you heard an argument (intentionally?) exclude the middle position when indeed there is a middle ground? Either you're with me or you're against me. Either you favor assisted suicide or you favor people suffering a lingering death. America, love it or leave it. These are not instances of the excluded middle; in a proper statement of the excluded middle, there is no in-between. Politicians frequently word their arguments as if the middle is excluded, forcing their opponents into positions they do not hold.

Interestingly enough, this black-and-white fallacy was common even among the politicians of ancient Greece. The Sophists, whom Plato and Aristotle dismissed with barely concealed contempt, attempted to use verbal maneuvering that sounded like the law of the excluded middle. For example, in Plato's *Euthydemus*, the Sophists convinced a young man to agree that he was either "wise or ignorant," offering no middle ground when indeed there should be.[2]

Closely related to the law of the excluded middle is the law of noncontradiction. The law of noncontradiction requires that a thing cannot both be and not be at the same time. A shape cannot be both a circle and not a circle. A figure cannot be both a square and not a square. Two lines in a plane cannot both intersect and not intersect. A statement cannot be both true and not true. When he developed his rules for logic, Aristotle repeatedly justified a statement by saying that it is impossible that "the same thing both is and is not at the same time."[3] Should you believe that a statement is both true and not true *at the same time*, then you find yourself mired in self-contradiction. A system of rules for proof would seek to prevent this. The Stoics, who developed further rules of logic in the third century B.C., acknowledged the law of the excluded middle and the law of noncontradiction in a single rule, "Either the first or not the first"—meaning always one or the other but never both.

The basic steps in any deductive proof, either mathematical or metaphysical, are the same. We begin with true (or agreed upon) statements, called *premises*, and concede at each step that the next statement or construction follows legitimately from the previous statements. When we arrive at the final statement, called our *conclusion*, we know it must necessarily be true due to our logical chain of reasoning.

Mathematics historian William Dunham asserts that although many other more ancient societies discovered mathematical properties through observation, the notion of *proving* a general mathematical result began with the Greeks. The earliest known mathematician is considered to be Thales who lived around 600 B.C.

A pseudo-mythical figure, Thales is described as the father of demonstrative mathematics whose legacy was his insistence that geometric results should not be accepted by virtue of their intuitive appeal, but rather must be "subjected to rigorous, logical proof."[4] The members of the mystical, philosophical, mathematical order founded in the sixth century B.C. by another semi-mythical figure, Pythagoras, are credited with the discovery and systematic proof of a number of geometric properties and are praised for insisting that geometric reasoning proceed according to careful deduction from *axioms*, or *postulates*. There is little question that they knew the general ideas of a deductive system, as did the members of the Platonic Academy.

There are numerous examples of Socrates' use of a deductive system in his philosophical arguments, as detailed in Plato's dialogues. Here we also bear witness to Socrates' use of the law of noncontradiction in his refutation of metaphysical arguments. Socrates accepts his opponent's premise as true, and by logical deduction, forces his opponent to accept a contradictory or absurd conclusion. What went wrong? If you concede the validity of the argument, then the initial premise must not have been true. This technique of refuting a hypothesis by baring its inconsistencies takes the following form: If statement ***P*** is true, then statement ***Q*** is true. But statement ***Q*** cannot be true. (***Q*** is absurd!) Therefore, statement ***P*** cannot be true. This form of argument by refutation is called *reductio ad absurdum*.

Although his mentor Socrates may have suggested this form of argument to Plato, Plato attributed it to Zeno of Elea (495–435 B.C.). Indeed, Aristotle gave Zeno credit for what is called *reductio ad impossibile*—getting the other to admit an impossibility or contradiction. Zeno established argument by refutation in philosophy and used this method to confound everyone when he created several paradoxes of the time, such as the well-known paradox of Achilles and the tortoise. The form of Zeno's argument proceeded like this: If statement ***P*** is true then statement ***Q*** is true. In addition, it can be shown that if statement ***P*** is true then statement ***Q*** is not true. Inasmuch as it is impossible that statement ***Q*** is both true and not true at the same time (law of noncontradiction), it is therefore impossible that statement ***P*** is true.[5]

Proof by Contradiction

Argument by refutation can prove only negative results (i.e., ***P*** is impossible). However, with the help of the double negative, one can prove all sorts of affirmative statements. Reductio ad absurdum can be used in proofs by assuming as *false* the statement to be proven. To prove an affirmative, we adopt as a premise the opposite of what we want to prove—namely, the contradictory of our conclusion. This way, once we have refuted the premise by an absurdity, we have proven that the opposite of what we wanted to prove is impossible. Today this is called an indirect proof or a proof by contradiction. The Stoics used this method to validate their rules of logic, and Euclid employed this technique as well.

While tangible evidence of the proofs of the Pythagoreans has not survived, the proofs of Euclid have. Long considered the culmination of all the geometry the Greeks knew at around 300 B.C. (and liberally borrowed from their predecessors), Euclid's *Elements* derived geometry in a thorough, organized, and logical fashion. As such, this system of deriving geometric principles logically from a few accepted postulates has become a paradigm for demonstrative proof. *Elements* set the standard of rigor for all of the mathematics that followed.[6]

Euclid used the method of "proof by contradiction" to prove

that there is an infinite number of prime numbers. To do this, he assumed as his initial premise that there is not an infinite number of prime numbers, but rather, that there is a finite number. Proceeding logically, Euclid reached a contradiction in a proof too involved to explain here. Therefore—what? What went wrong? If the logic is flawless, only the initial assumption can be wrong. By the law of the excluded middle, either there is a finite number of primes or there is not. Euclid, assuming that there was a finite number, arrived at a contradiction. Therefore, his initial premise that there was a finite number of primes must be false. If it is false that "there is a finite number of primes" then it is true that "there is not a finite number." In other words, there is an infinite number.

Euclid used this same technique to prove the theorem in geometry about the congruence of alternate interior angles formed by a straight line falling on parallel lines (Fig. 3). To prove this proposition, he began by assuming that the alternate interior angles formed by a line crossing parallel lines are *not* congruent (the same size) and methodically proceeded step by logical step until he arrived at a contradiction. This contradiction forced Euclid to conclude that the initial premise must be wrong and therefore alternate interior angles are congruent.

To use the method of proof by contradiction, one assumes as a premise the opposite of the conclusion. Oftentimes figuring out the opposite of a conclusion is easy, but sometimes it is not. Likewise, to refute an opponent's position in a philosophical

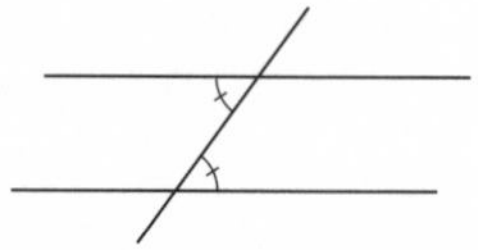

Figure 3. One of the geometry propositions that Euclid proved: Alternate interior angles must be congruent.

argument, we need to have a clear idea of what it means to contradict his position. Ancient Greek debates were carried out with two speakers holding opposite positions. So, it became necessary to understand what contradictory statements were to know at what point one speaker had successfully refuted his opponent's position. Aristotle defined statements that *contradict* one another, or statements that are in a sense "opposites" of one another. Statements such as "No individuals are altruistic" and "Some individual(s) is (are) altruistic" are said to be *contradictories*. As contradictories, they cannot both be true and cannot both be false—one must be true and the other false.

Aristotle declared that every affirmative statement has its own opposite negative just as every negative statement has an affirmative opposite. He offered the following pairs of contradictories as illustrations of his definition.

Aristotle's Contradictory Pairs[7]

It may be	It cannot be.
It is contingent [uncertain].	It is not contingent.
It is impossible.	It is not impossible.
It is necessary [inevitable].	It is not necessary.
It is true.	It is not true.

Furthermore, a statement such as "Every person has enough to eat" is *universal* in nature, that is, it is a statement about all persons. Its contradictory statement "Not every person has enough to eat" or "Some persons do not have enough to eat" is not a universal. It is said to be *particular* in nature. Universal affirmations and particular denials are contradictory statements.

Likewise, universal denials and particular affirmations are contradictories. "No individuals are altruistic" is a universal denial, but its contradiction, "Some individuals are altruistic," is a particular affirmation. As contradictories, they cannot both be true and cannot both be false—it will always be the case that one statement is true and the other is false.

Individuals often confuse contradictories with contraries. Aristotle defined *contraries* as pairs of statements—one affirmative and the other negative—that are both universal (or both particular) in nature. For example, "All people are rich" and "No people are rich" are contraries. Both cannot be true yet it is possible that neither is true (that is, both are false).

> "No one in this family helps out." "Some of us help out." "Don't contradict me."
>
> "Everyone in this family is lazy." "I hate to contradict you, but some of us are not lazy."
>
> "No one in this family helps out." "We all help out." "Don't be contrary."
>
> "Everyone in this family is lazy." "To the contrary, none of us is lazy."

John Stuart Mill noted the frequent error committed when one is unable to distinguish the *contrary* from the *contradictory*.[8] He went on to claim that these errors occur more often in our private thoughts—saying that if the statement were enunciated aloud, the error would in fact be detected.

Disproof

Disproof is often easier than proof. Any claim that something is absolute or pertains to all of something needs only one *counterexample* to bring the claim down. The cynic asserts, "No human being is altruistic." If you can think of one human being who has ever lived who is altruistic, you can defeat the claim. For example, you might get the cynic to admit, "Mother Teresa is altruistic." Therefore, some human being is altruistic and you have brought down the cynic's claim with one counterexample. As Albert Einstein suggested, any number of instances will never prove an "all" statement to be true, but it takes a single example to prove such a statement false.

In the face of an "all" or "never" statement, one counterexample can disprove the statement. However, in ordinary discourse we frequently hear the idea of a counterexample being used incorrectly. The idea of argument by counterexample does not extend in the reverse direction. Nonetheless, we sometimes hear the illogic that follows: She: All women are pacifists. He: I'm not a woman and I'm a pacifist. (This is not a counterexample. To disprove her statement, he must produce a woman who is not a pacifist.)

Psychologists have found that people can be extremely logical when they can notice a contradiction but that correct inference is often hindered when a counterexample is not obvious. For example, in Guy Politzer's study on differences in interpretation of the logical concept called the *conditional*, his subjects were highly successful in evaluating a rule logically when direct evidence of a contradiction was present. Specifically, Politzer's subjects were given a certain statement such as, "I never wear my dress without wearing my hat," accompanied by four pictures similar to those in Figure 4. Subjects were asked to label each picture as "compatible" or "incompatible" with the given statement. Inasmuch as the pictures illustrated the only possible combinations of information, subjects weren't required to retrieve that information from memory. These visual referents facilitated the retrieval of a contradiction.[9]

Examine the pictures in the figure for yourself. From left to right, they illustrate hat/dress, no hat/dress, hat/no dress, and no hat/no dress. The claim is made, "I never wear my dress without wearing my hat," and we are to judge whether the pictures are consistent or inconsistent with the claim. Since the claim is about what I will or will not do when I wear my dress, we judge that the last two pictures are "compatible" with the claim as they are not inconsistent with it. The first two pictures must be examined in more detail since the wearing of a dress is directly addressed by the claim. "I never wear my dress without wearing my hat" is clearly consistent with the first picture and is clearly violated by the second. So the correct answers are that all the pictures are "compatible" with the claim except the second, which is "incompatible" with it.

In this experiment, subjects were not obliged to rely on memory or imagine all possible dress/hat scenarios. The subjects were presented with pictorial reminders of every possibility. With visual images at hand, subjects could label those pictures that contradicted the statement as incompatible; otherwise the pictures were compatible.

From very ancient times, scientists have sought to establish

Figure 4. Evaluate each picture as compatible or incompatible with the statement "I never wear my dress without wearing my hat."

universal truths, and under the influence of Thales, Pythagoras, and Euclid, universal truths required proof. Armed with the law of the excluded middle and the law of noncontradiction, ancient mathematicians and philosophers were ready to deliver *proof*. All that remained was an agreed-upon set of rules for logical deduction. Aristotle and the Stoics provided such a framework for deductive inference, and the basics of their systems remain virtually unchanged to this day.

As the Greek philosophers attempted to establish universal truths about humans and the world around them, definitions were set forth in an effort to find a common ground in language. Aristotle defined statements of truth or falsity and words like *all*. Do they really need any definition? He felt that for one to articulate a system of correct thinking, nothing should be taken for granted. As we'll see in the next chapter, he was right.

2
ALL

You may fool all the people some of the time;
you can even fool some of the people all the time;
but you can't fool all of the people all the time.
Abraham Lincoln

Aristotle's works in logic consisted of six treatises: *Categories, On Interpretation*, *Prior Analytics* (or *Concerning Syllogism*), *Posterior Analytics* (or *Concerning Demonstration*), *Topics*, and *Sophistical Elenchi* (or *On Sophistical Refutations*). After Aristotle's death in 322 B.C., his followers collected these treatises into a single work known as the *Organon*, or instrument of science.

The title, *On Interpretation*, reflects the notion that logic was regarded as the interpretation of thought.[1] In this treatise, Aristotle set down rules of logic dealing with statements called *propositions*. A proposition is any statement that has the property of truth or falsity. A prayer, Aristotle says, is not a proposition. "Come here" and "Where are you?" are not propositions. "2 + 2 = 5" is a proposition (it is false). "Socrates was a man" is a proposition (it is true). Propositions can be true or false and nothing in between (law of the excluded middle), but not both true and false at the same time (law of noncontradiction).[2] "All tornadoes are destructive" might be a false proposition if it is true that some tornadoes are not destructive, even if only one is not.

"That tornado is destructive" would certainly be either true or false but not both. We would know whether the proposition is true or false by checking the facts and agreeing on a definition of "destructive." "Some tornadoes are destructive" would qualify as a proposition, and we would all probably agree it is a true proposition, having heard of at least one tornado that met our definition of "destructive."

Terms called *quantifiers* are available for making propositions. Quantifiers are words such as *every*, *all*, *some*, *none*, *many*, and *few*, to name a few. These words allow a partial quantification of items to be specified. Although words like *some*, *many*, and *few* may provide only a vague quantification (we don't know how many *many* is), words like *all* and *none* are quite specific.

The English words *all* and *every* are called (affirmative) *universal quantifiers* in logic. They indicate the totality (100 percent) of something. Sometimes the *all* is implied, as in "Members in good standing may vote." However, if we want to emphasize the point, we may say, "All persons are treated equally under the law." The word *any* is sometimes regarded as a universal quantifier. "Any person who can show just cause why this man and woman should not be joined in holy wedlock. . . ." The article *a* may also be used as a universal quantifier, as in "A library is a place to borrow books" meaning "All libraries are places to borrow books." Universal affirmative propositions such as these were called *de omni*, meaning *all*, by Latin commentators on Aristotle.

It has been shown that the universality of the word *all* is clearer than the universality of *any* and *a*. In a 1989 study, David O'Brien and his colleagues assessed the difficulty of different formulations of the universal *all* by testing second graders, fourth graders, eighth graders, and adults.[3] Without exception, in every age group the tendency to err was greatest when the indefinite article *a* was used, "If *a* thing. . . ." For older children and adults, errors decreased when *any* was used, "If *any* thing . . . ," and errors virtually vanished when the universality was made explicit, "*all* things. . . ." With the youngest children, though the errors did not vanish, they were reduced significantly when the universality was made clear with the word *all*.

All *S* are *P*

In addition to a quantifier, each proposition contains a *subject* and a *predicate*. For example, in the universal affirmation "All men are human beings," the class of *men* is called the subject of the universal proposition and the class of *human beings* is called the predicate. Consequently, in logic books, the universal affirmation is often introduced to the reader as "All ***S*** are ***P***."

Although not truly an "all" statement, one other type of proposition is classified as a universal affirmation: "Socrates was a Greek." "I am a teacher." These propositions do not, on the surface, appear to be universal propositions. They are called *singular* or *individual* and are treated as universal claims. Even though the statements speak of a single individual, they are interpreted as constituting an entire class that has only a single entity in it.[4] Classical logic construes the propositions as, "All things that are identical with Socrates were Greek" or "All things that belong to the class of things that are me are teachers."

Vice Versa

Given the right example, it is clear that the statement "All ***S*** are ***P***" is not the same as the statement "All ***P*** are ***S***." We would probably agree "All mothers are parents" is a true statement whereas "All parents are mothers" is not. Yet this *conversion* is a common mistake. These two statements, "All ***S*** are ***P***" and "All ***P*** are ***S***," are called *converse* statements. They do not mean the same thing. It is possible that one is true and the other is not. It is also possible that both are true or neither is true. You might think of the converse as the *vice versa*. All faculty members are employees of the university, but not vice versa. All dogs love their owners and vice versa. (Although I'm not sure either is true.)

According to Bärbel Inhelder and Jean Piaget, children aged 5 and 6 have trouble with the quantifier *all* even when information is graphic and visual. In their experiments, they laid out red square counters and blue circle counters, adding some blue squares, all of which the children were allowed to see during their questioning. Using white and gray counters, their experiment involved a set of objects such as those in Figure 5. Children were then asked questions such as "Are all the squares white?" (NO) and "Are all the circles gray?" (YES.) More difficult for the younger children were questions such as "Are all the white ones squares?" (YES.) The youngest subjects converted the quantification 50 percent of the time, thinking that "All the squares are white" meant the same as "All the white ones are squares."[5] This may be explained in part by the less developed language ability of the youngest children, but their mistakes may also be explained by their inability to focus their attention on the relevant information.

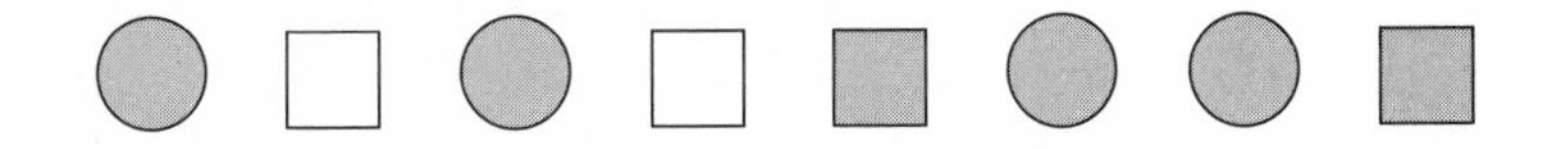

Figure 5. Which statements are true?
"All squares are white. All white things are squares."
"All circles are gray. All gray things are circles."

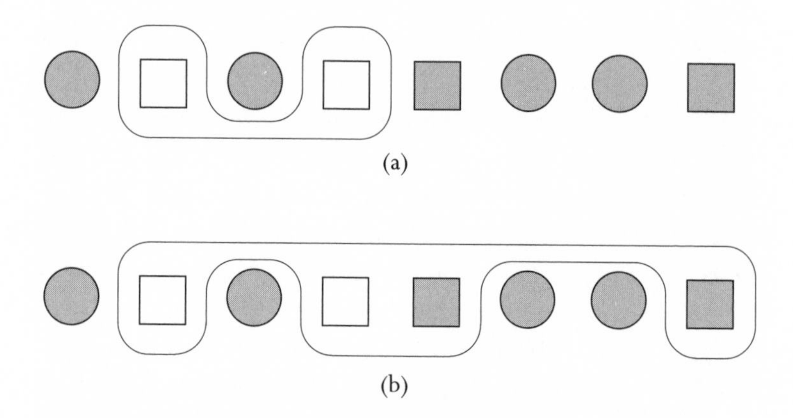

Figure 6. (a) Are all the white things squares? (b) Are all the squares white? To correctly answer these questions, we must focus our attention on the pertinent information.

Inhelder and Piaget noted the difficulty of mastering the idea of class inclusion in the youngest children (Fig. 6). That is, the class of white squares is included in the class of squares, but not vice versa. By ages 8 and 9, children were able to correctly answer the easier questions 100 percent of the time and produced the incorrect conversion on the more difficult questions only 10 to 20 percent of the time.

Understanding the idea of class inclusion is important to understanding "all" propositions. If the statement "All taxicabs are yellow" is true, then the class of all taxicabs belongs to the class of all yellow cars. Or, we could say that the set of all taxicabs is a subset of the set of all yellow cars. Sometimes a visual representation like Figure 7 is helpful, and quite often diagrams are used as illustrative devices.

The introduction of diagrams to illustrate or solve problems in logic is usually attributed to the brilliant Swiss mathematician Leonhard Euler. His diagrams were contained in a series of let-

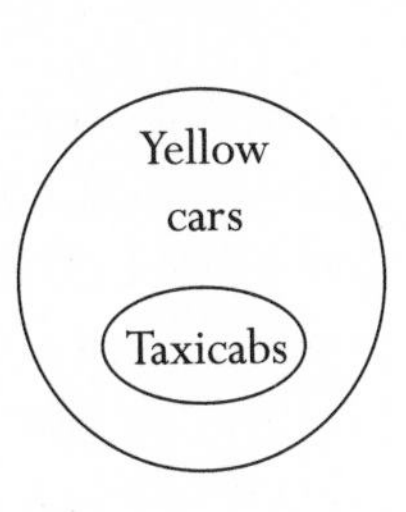

Figure 7. Graphic representation of "All taxicabs are yellow."

ters written in 1761 to the Princess of Anhalt-Dessau, the niece of Frederick the Great, King of Prussia. The famous *Letters to a German Princess (Lettres à une Princesse D'Allemagne)* were published in 1768, proved to be immensely popular, and were circulated in book form in seven languages.[6] Euler's letters were intended to give lessons to the princess in mechanics, physical optics, astronomy, sound, and several topics in philosophy, including logic. One translator, writing in 1795, remarked on how unusual it was that a young woman of the time had wished to be educated in the sciences and philosophy when most young women of even the late eighteenth century were encouraged to learn little more than the likes of needlepoint.[7]

Euler's instruction in logic is not original; rather, it is a summary of classical Aristotelian and limited Stoic logic. It turns out that his use of diagrams is not original either. The identical diagrams that the mathematical community called *Euler's circles* had been demonstrated earlier by the German "universal genius" Gottfried Leibniz. A master at law, philosophy, religion, history, and statecraft, Leibniz was two centuries ahead of his time in logic and mathematics. Most of his work in logic was not published until the late nineteenth century or early twentieth century, but around 1686 (one hundred years before the publication of Euler's famous *Letters*), Leibniz wrote a paper called *De Formae Logicae Comprobatione per Linearum Ductus*, which contained the figures that became known as Euler's circles. The diagrams are one and the same; there is no way that Euler could not have seen them previously. Most likely, the idea had been suggested to him through his mathematics tutor, Johann Bernoulli. The famous Swiss mathematicians, brothers Jakob and Johann Bernoulli, had been avid followers of Leibniz and disseminated his work throughout Europe.

Although his mathematical ability is legendary, Euler was also noted for his ability to convey mathematical ideas with great clarity. In other words, he was an excellent teacher. Like any good teacher, he used any device in his repertoire to instruct his students. Euler's impact on the mathematical world was so influential that his style and notation were often imitated. Thus, the idea of using diagrams in logic was assigned to him.

The Leibniz/Euler circles exhibit the proposition "Every **A** is **B**" in the same way we earlier displayed "All taxicabs are yellow"—with the class of **A**-things represented as a circle inside the circle of **B**-things. Perhaps more familiar to the reader, and widely considered an improvement on the Leibniz/Euler circles, is the Venn diagram.[8]

John Venn, the English logician and lecturer at Cambridge University, first published his method of diagrams in an 1880 *Philosophical Magazine* article, "On the Diagrammatic and Mechanical Representation of Propositions and Reasoning." Venn would have represented "All taxicabs are yellow" with two overlapping circles as shown in Figure 8, shading the portion of the taxicab circle that is outside the yellow-cars circle as an indication that there is nothing there. The shaded portion indicates that the class of non-yellow taxicabs is empty.

At first glance, Venn's diagram does not seem as illustrative as the Leibniz/Euler diagram—their diagram actually depicts the class of taxicabs *inside* the class of yellow cars. However, as we will

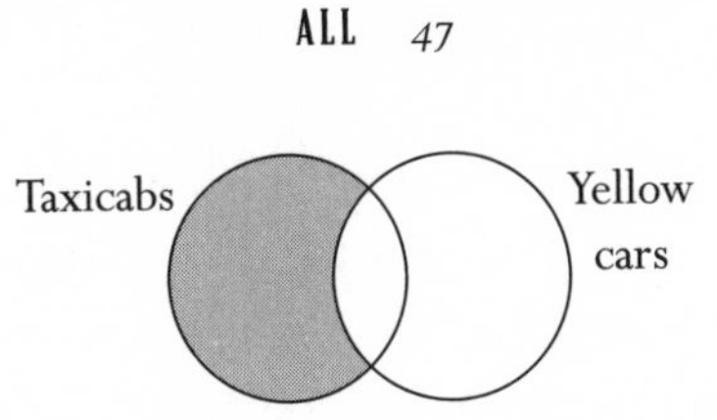

Figure 8. A Venn diagram of "All taxicabs are yellow."

later see, Venn's diagram has the advantage of being much more flexible. Many other philosophers and mathematicians have devised diagrammatic techniques as tools for analyzing propositions in logic. The American scientist and logician Charles Sanders Peirce (pronounced "purse") invented a system comparable to Venn's for analyzing more complicated propositions. Lewis Carroll devised a system resembling John Venn's—using overlapping rectangles instead of circles—and used an O to indicate an empty cell, as in Figure 9. Both Peirce and Carroll were huge advocates of teaching logic to schoolchildren through the use of graphs such as these. Educators must have been paying attention, because schoolchildren today are taught classification skills from a very early age by the use of Venn's overlapping circles.

Euler also found the figures valuable as a teaching tool. He

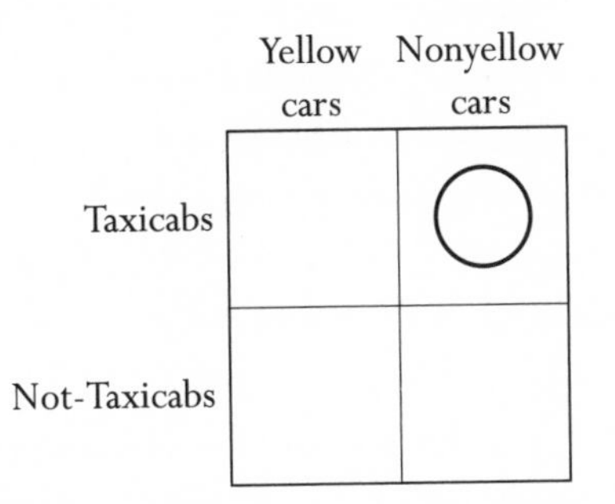

Figure 9. "All taxicabs are yellow," in the style of Lewis Carroll.

noted that the propositions in logic may "be represented by figures, so as to exhibit their nature to the eye. This must be a great assistance, toward comprehending, more distinctly, wherein the accuracy of a chain of reasoning consists."[9] Euler wrote to the princess,

> These circles, or rather these spaces, for it is of no importance what figure they are of, are extremely commodious for facilitating our reflections on this subject, and for unfolding all the boasted mysteries of logic, which that art finds it so difficult to explain; whereas, by means of these signs, the whole is rendered sensible to the eye.[10]

It is interesting that in 1761 Euler mentions the difficulty of explaining the art of logic. This fact should be of some comfort to teachers everywhere. Even today, instructors at the university level see these misunderstandings crop up in math, philosophy, and computer science classes time after time. While adults would probably have little difficulty dealing with Inhelder and Piaget's questions with colored counters, when the information is presented abstractly, without a visual referent, even adults are likely to reach the wrong conclusion from a given set of statements. Yet, according to Inhelder and Piaget, by approximately the twelfth grade, most of us have reached our formal reasoning period and should have the ability to reason logically.

Familiarity—Help or Hindrance?

Unlike the visual clues provided in Inhelder and Piaget's study of logical reasoning in children or the pictures provided in Politzer's study as mentioned earlier, we are usually required to reason without access to direct evidence. Without evidence at hand, we must recall information that is often remote and vague. Sometimes our memory provides us with counterexamples to prevent our faulty reasoning, but just as often our memory leads us astray.

The rules of inference dictating how one statement can follow from another and lead to logical conclusions are the same regardless of the content of the argument. Logical reasoning is supposed to take place without regard to either the sense or the truth of the statement or the material being reasoned about. Yet, often reasoning is more difficult if the material under consideration is obscure or alien. As one researcher put it, "The difficulty of applying a principle of reasoning increases as the meaningfulness of the content decreases."[11] The more abstract or unfamiliar the material, the more difficult it is for us to draw correct inferences.

In one of the earliest studies examining the content or material being reasoned about, M. C. Wilkins in 1928 found that when given the premise, "All freshmen take History I," only 8 percent of her subjects erroneously accepted the conversion, "All students taking History I are freshmen." However, 20 percent of them accepted the equally erroneous conclusion, "Some students taking History I are not freshmen." With strictly symbolic material (All ***S*** are ***P***), the errors "All ***P*** are ***S***" and "Some ***P*** are not ***S***" were made by 25 percent and 14 percent of the subjects, respectively. One might guess that in the first instance students retrieved common knowledge about their world—given the fact that all freshmen take History I does not mean that only freshmen take it. In fact, they may have themselves observed nonfreshmen taking History I. So their conclusion was correct and they were able to construct a counterexample to prevent making the erroneous conversion. However, as they continued thinking along those lines, knowledge about their own world encouraged them to draw a (possibly true) conclusion that was not based on correct logical inference. "Some students taking History I are not freshmen" may or may not be true, but it does not logically follow from "All freshmen take History I." Interestingly enough, when abstract material was used and subjects could not tap into their own experience and knowledge about the material, more of them made the conversion mistake (for which there are countless concrete examples that one can retrieve from memory—"All women are human" doesn't mean "All humans are women") while fewer made the second inference mistake.

"All horn players have good chops." My husband, a singer extraordinaire, can see right through this trap. He will not accept the converse statement "All people with good chops play the horn." He's not a horn player but he does have good chops. With evidence at hand he avoids the common fallacy because he recognizes a counterexample or inconsistency in accepting the faulty conclusion.

Clarity or Brevity?

There seem to be two different systems of language—one is that of natural language and the other that of logic. Often the information we convey is the *least* amount necessary to get our points across.

Dr. Susanna Epp of De Paul University uses the example of a classroom teacher who announces, "All those who sit quietly during the test may go outside and play afterward."[12] Perhaps this is exactly what the teacher means to say. And, if so, then she means that those who will get to go out and play will definitely include the quiet sitters, but might well include those who make noise. In fact her statement says nothing at all about the noise-makers one way or the other. I doubt that the students interpret her this way.

Is the teacher intentionally deceiving the students? Is she hoping that students will misconstrue the statement? Chances are good that most of the students believe she is actually making the converse statement that all those who make noise will not get to play outside. Had the teacher made the statement "All those who do not sit quietly during the test may not go out and play afterwards," then the warning doesn't address the question of what will happen to the quiet sitters. She probably means, "All those who sit quietly during the test may go outside and play afterwards, and those who don't sit quietly may not go outside and play afterwards." In the interest of brevity, we must often take the speaker's meaning from the context of his or her language and our own life experiences.

Since logic defines strict rules of inference without regard to content, we may be forced to accept nonsensical statements as true due to their correct form. How is one to evaluate the truth of "All my Ferraris are red" if I have no Ferraris? In ordinary language, we might say that it is neither true nor false—or that it is nonsense. Yet, the classical rules of logic require propositions to be either true or not true (law of the excluded middle). Some logicians have ignored this kind of proposition. They have made an *existential assumption*, that is, an assumption that the subject of any universal proposition exists. Others make no existential assumption, claiming that the diagrams of Leibniz/Euler and Venn serve us well to represent the universal proposition regardless of whether the class of my Ferraris has any members or not. "All angels are good" and "All devils are evil" can be allowed as true propositions whether or not angels or devils exist.[13]

Of course, things could get much more complicated. We have

only considered universal quantifiers and have only quantified the subject of the proposition. In ordinary language, we put quantifiers anywhere we want. And what if we put the word "not" in front of "all"? ***Not** all* drastically changes the proposition, not only changing it from an affirmation to a negation but also changing its universal nature. Even when the rules of logic were being developed, Aristotle recognized that *negation* makes reasoning a good deal more difficult. So naturally he addressed rules of negation. Let's examine them next.

3
A NOT Tangles Everything Up

"No" is only "yes" to a different question.
Bob Patterson

If every instinct you have is wrong,
then the opposite would have to be right.
Jerry Seinfeld

We encountered negations very early on while examining the law of the excluded middle and the law of noncontradiction. While Aristotle reminded us that it is impossible that the same thing both *is* and *is not* at the same time, he also recognized that we can construct both an affirmation and a negation that have identical meanings. Aristotle said that there are two types of propositions that are called *simple*—the *affirmation*, which is an assertion, and the *negation*, which is a negative assertion or a denial. All others are merely conjunctions of simple propositions.

"All humans are imperfect" is an affirmation, while "No human is perfect" is a denial with the same meaning. "Tuesday you were absent" is an affirmation, and "Tuesday you were not present" is a denial conveying the same information. "Four is not an odd number" is a true negation and "four is an even number" is a true affirmation expressing the same information from a different perspective. Inasmuch as it is possible to affirm the absence of something or to deny the presence of something, the same set of facts may be stated in either the affirmative or the negative.

So what does the negation of an "all" statement look like? Consider the negation of a simple sentence such as "All the children like ice cream." Its negation might well read, "It is not the case that all the children like ice cream." But even long ago Aristotle suggested that the negation be posed as the contradictory statement, such as "Not every child likes ice cream" or "Some children don't like ice cream." We could negate using the passive voice—"Ice cream isn't liked by every child" or "Ice cream isn't liked by some of the children." The underlying structure of any of these negations is simply ***not***-(all the children like ice cream).

The Trouble with *Not*

The noted logic historians William and Martha Kneale state that from the time of Parmenides in the fifth century B.C., the Greeks found something mysterious in negation, perhaps associating it with falsehood.[1] In modern times, some researchers have argued that negation is not "natural" since it is hardly informative to know what something is not. However, more often than we may realize the only way to understand what something *is* is to have a clear understanding of what it *isn't*. How would we define an odd number other than by saying it is a number that is not divisible by 2? What is peace but the absence of war?

Another argument put forth relative to the difficulty of reasoning with negation concerns the emotional factor. This position argues that the prohibitive nature of words such as "no" and "not" makes us uncomfortable. Some psychologists have suggested that since negation is fraught with psychological problems, negation necessarily increases the difficulty inherent in making inferences.[2]

Cognitive psychologists Peter C. Wason and Philip Johnson-Laird have written several books and dozens of articles on how we reason. They point out that negation is a fundamental concept in reasoning, a concept so basic to our everyday thinking that no known language is without its negative terms.[3] Negation ought to be an easy, perhaps the easiest, form of deduction. However, making even a simple inference involving a negative is a two-step process. If I say, "I am not an ornithologist," two statements must be absorbed. First, we must grasp what it means to be an ornithologist, then what it means not to be one. In our day-to-day communication, the extra step involved in reasoning with negation may well go unnoticed.

In one of their studies, Wason and Johnson-Laird performed a series of experiments focusing on the reasoning difficulties associated with negation. When asked questions that involved affirmation and negation, their subjects were slower in evaluating the truth of a negation than the falsity of an affirmation and got it wrong more often—a clear indication that negation is a more difficult concept to grasp.[4]

Negation may be either implicit or explicit. There is evidence that in some instances an *implicit negative* is easier to correctly process than an *explicit negative*. Implicit negatives are words that have negative meaning without using the word "not." Implicit negatives, such as "absent" rather than "not present," "reject" rather than "not accept," and "fail" rather than "not pass," may be easier to deal with than their explicitly negative counterparts. In other instances, implicit negatives may be too well hidden. For example, researchers have indicated that it is easier to see that the explicit negative, "The number is not 4," negates "The number is 4" but more difficult to see that the implicit negative, "The number is 9," also negates "The number is 4."[5]

Researcher Sheila Jones tested the ease with which differently worded instructions were handled by individuals. Three sets of directions were tested that all had the same meaning—one set of instructions was an affirmative, one a negative, and one an implicit negative.[6] The subjects were presented a list of digits, 1 through 8, and given one of the following sets of instructions:

Mark the numbers 1, 3, 4, 6, 7. (affirmative)

Do not mark the numbers 2, 5, 8, mark all the rest. (negative)

Mark all the numbers except 2, 5, 8. (implicit negative)

The test was set up in a manner similar to that shown in Figure 10. The subjects' speed and accuracy were measured as indicators of difficulty. The subjects performed the task faster and with fewer errors of omission following the affirmative instruction even though the list of numbers was considerably longer. Subjects performing the task using "except" were clearly faster than those following the "not" instruction, signifying that the implicit negatives were easier to understand than the instructions containing the word "not."

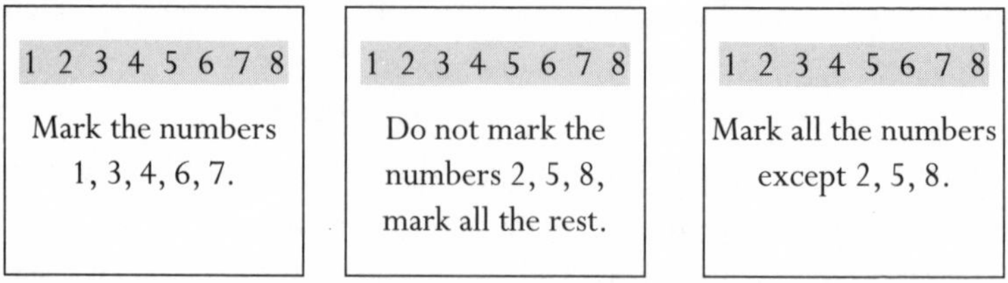

1 2 3 4 5 6 7 8	1 2 3 4 5 6 7 8	1 2 3 4 5 6 7 8
Mark the numbers 1, 3, 4, 6, 7.	Do not mark the numbers 2, 5, 8, mark all the rest.	Mark all the numbers except 2, 5, 8.

Figure 10. Task measuring the difficulty of the affirmative, negative, and implicit negative.

Some negatives do not have an implicit negative counterpart, and those negatives are more difficult to evaluate. The statement "The dress is not red" is harder to process than a statement like "Seven is not even," because the negation "not even" can be easily exchanged for the affirmative "odd," but "not red" is not easily translated. "Not red" is also very difficult to visualize. The difficulties involved with trying to visualize something that is not may well interfere with one's ability to reason with negatives. If I say that I did not come by car, what do you see in your mind's eye?

It may be that, wherever possible, we translate negatives into affirmatives to more easily process information. To make this translation an individual must first construct a *contrast class*, like the class of not-red dresses or the class of modes of transportation that are not-car. The size of the contrast class and the ease with which a contrast class can be constructed have been shown to affect our ability to reason with negatives.[7]

Wason and Johnson-Laird suggest that in everyday language a denial often serves as a device to correct a preconceived notion. Although it is true that I am not an ornithologist, I am not likely to make that statement unless someone was under the misconception that I was. The statement "Class wasn't boring today" would probably not be made if the class were generally not boring. This kind of statement is usually made when the class is frequently or almost always boring. The statement functions to correct the listener's previously held impression by pointing out an exception.

An experiment by Susan Carry indicated that negatives used on an exceptional case were easier than negatives used on unexceptional cases. In her experiment, individuals were exposed to and then questioned about an array of circles, numbered 1 through 8. All of the circles except one were the same color, and the circle of exceptional color varied in its position number. Presumably, most of us would remember the array of circles by remembering the exceptional circle since this requires retaining the least amount of information. Her experiment confirmed that it is easier to negate an exceptional case in terms of the property that makes it exceptional than to negate the majority cases in terms of the property of the exception.[8]

In addition, the results of a study by Judith Greene showed that negatives used to change meaning were processed more easily than negatives used to preserve meaning. Subjects were asked to determine whether two abstract sentences had the same or different meanings. A series of tasks paired sentences sometimes with the same meaning, one involving a negation and the other not, and other times paired sentences with different meanings, one involving a negation and the other not. Greene labeled a negative that signified a change in meaning *natural*, while a negative that preserved meaning was dubbed *unnatural*. For example, "*x* exceeds *y*" and "*x* does not exceed *y*" are easily processed by the brain as being *different* in meaning (thus the negative is performing its *natural* function), while "*x* exceeds *y*" and "*y* does not exceed *x*" are more difficult to assess as having the *same* meaning. Her studies support the notion that we more easily digest negatives that change a preconception rather than negatives that confirm a previously held notion.[9]

Scope of the Negative

Aristotle went to great lengths in his treatises to point out that the negation of "All men are just" is the contradictory "It is not the case that all men are just," rather than the contrary "No men are just." In the negation, "It is not the case that all men are just," the scope of the negative is the entire assertion, "all men are just." The scope of the negative in the contrary "No men are just" is simply "men." The difference between the contradictory and the contrary is that the contradictory is the negation of an entire proposition and that is why the proposition and its contradictory are always opposite in truth value. When one is true, the other is false, and vice versa.

Aristotle recommended that the statement "It is not the case that all men are just" was more naturally communicated as "Some men are not just." Several studies have borne out the fact that this form may indeed be more natural. The smaller the scope of the negative, the easier the statement is to understand. Studies have shown that it takes systematically longer to process the type of denial involving "It is not the case that . . ." and "It is false that . . ." than ordinary negation. Indications are that statements where the scope of the negative is small, like "Some people do not like all ice cream flavors," are easier to process than ones such as "It is not the case that all people like all ice creams flavors."[10]

A and E Propositions

Medieval scholars of logic invented schemes and labels that became common terminology for students studying Aristotle's classification of propositions. The universal affirmation, "All ***S*** are ***P***," was named a type-**A** proposition. The universal negation or denial, "No ***S*** are ***P***," was named a type-**E** proposition. This pair of **A** and **E** statements are the *contrary* statements. As such, they cannot both be true, but exactly one could be true or both could be false. The type-**A** universal affirmation, "All people are honest in completing their tax forms," and the type-**E** universal

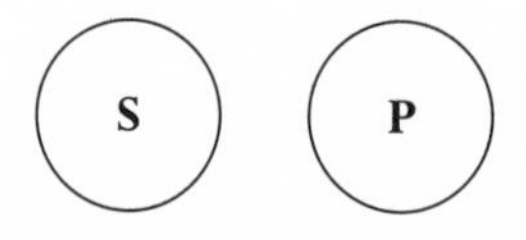

Figure 11. A Leibniz/Euler diagram of "No *S* are *P*."

denial, "No people are honest in completing their tax forms," are contraries. In this case, both are probably false.

The Leibniz/Euler logic diagrams represent the universal negation, "No ***S*** are ***P***," as two spaces separate from each other—an indication that nothing in notion ***S*** is in notion ***P***. The proposition, "No ***S*** are ***P***," is seen in Figure 11.

John Venn's diagrams once again employed the use of overlapping circles to denote the subject and the predicate. In fact, all of Venn's diagrams use the overlapping circles, which is one of its most attractive features. Using Venn's graphical method, all of the Aristotelian propositions can be represented by different shadings of the same diagram—using one piece of graph paper, so to speak. Again, Venn's shaded region indicates emptiness—nothing exists there. So in representing "No ***S*** are ***P***," the region where ***S*** and ***P*** overlap is shaded to indicate that nothing can be there, as shown in Figure 12.

Earlier, we witnessed the error in logic called conversion that is commonly made with the universal affirmative (type-**A**) proposition. It is a mistake to think "All ***S*** are ***P***" means the same thing as

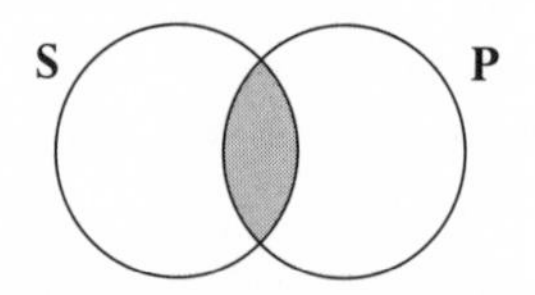

Figure 12. A Venn diagram of "No *S* are *P*."

"All ***P*** are ***S***." Quite frequently one is true and the other is not. Just because all zebras are mammals doesn't mean that all mammals are zebras. Yet, *converting* a type-**E** proposition (a universal negation) is not an error. "No chickens are mammals" and "no mammals are chickens" are both true. In fact, any time "No ***S*** are ***P***" is true, so is "No ***P*** are ***S***." This fact becomes crystal clear by looking at either the Leibniz/Euler diagram or the Venn diagram. In the Leibniz/Euler diagram, nothing in space ***S*** is in space ***P*** and nothing in space ***P*** is in space ***S***. In John Venn's diagram, nothing ***S*** is in ***P*** and nothing ***P*** is in ***S***. Imagine what the diagrams for "No ***P*** are ***S***" would look like. Using either diagram, it is clear that the figure for "No ***P*** are ***S***" would look exactly the same as "No ***S*** are ***P***" with perhaps the labels on the circles interchanged.

When *No* Means *Yes* —The "Negative Pregnant" and Double Negative

In his *On Language* column, William Safire discussed a fascinating legal term called the *negative pregnant* derived from fifteenth-century logicians.[11] The *Oxford English Dictionary* notes that a *negative pregnant* means "a negative implying or involving an affirmative." If asked, "Did you steal the car on November 4?" the defendant replying with the negative pregnant "I did not steal it on November 4" leaves the possibility (maybe even the implication) wide open that he nonetheless stole the car on some day. Early on in life, young children seem to master this form of avoiding the issue. When asked, "Did you eat the last cookie yesterday?" we might well hear, "I did not eat it yesterday" or, "Yesterday? . . . No."

Double negatives fascinate us from the time we first encounter them in elementary school. They cropped up earlier in the discussion of proofs by contradiction, where we begin by assuming the opposite of that which we want to prove. If I want to prove proposition ***P***, I assume ***not-P***. Proceeding by impeccable logic, I arrive at a contradiction, an impossibility, something like 0 = 1. What went wrong? My initial assumption must be false. I conclude, "***not-P*** is false" or "it is not the case that ***not-P***" or "not-***not-P***." The equivalence of the statements "not (***not-P***)" and "***P***"—that the negation of a negation yields a affirmation—was a principle in logic recognized by the Stoics as early as the second century B.C.[12]

All too frequently for the electorate we see double negatives in referendum questions in the voting booth. This yes-means-no and no-means-yes wording is often found in propositions to repeal a ban on something. A vote "yes" on the repeal of term limits means you do not favor term limits. A vote "no" on the repeal of the ban on smoking means you favor smoking restrictions. A vote "no" to repeal a ban on gay marriages means you favor restrictions on gay marriages, but a "yes" vote to repeal the ban on assault weapons means you do not favor restrictions on assault weapons. I recently received a ballot to vote for some proposals in the management of my retirement funds. The ballot question is in Figure 13. If you are like people who want their money invested in issues they favor (some folks don't care), voting "for" means you are against gun control and voting "against" means you favor gun control.

Proposal: To stop investing in companies supporting gun control.

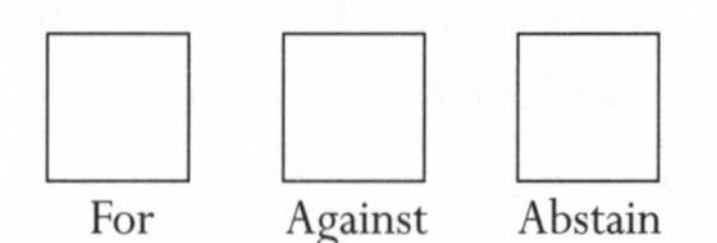

Figure 13. Example of when voting "for" means against.

Studies have shown that reasoners find it difficult to negate a negative.[13] If the process of negation involves an extra mental step, a double negative can be mind boggling. Statements such as "The probability of a false negative for the pregnancy test is 1 percent" or "No non-New Yorkers are required to complete form 203" or "The statistical test indicates that you cannot reject the hypothesis of no difference" can cause listeners to scratch their heads (or give them a headache).

As we mentioned earlier, a statement like "It is not the case that all men are honest" is more naturally communicated as "Some men are not honest." But *some* is not universal. So Aristotle defined propositions dealing with *some are* and *some are not*. Do they really need definition? You may be surprised to learn that they mean different things to different people. Read on.

4

SOME Is Part or All of ALL

If every boy likes some girl and every girl likes some boy, does every boy like someone who likes him?

JONATHAN BARON,
Thinking and Deciding

Although statements about "all" of something or "none" of something are powerful and yield universal laws in mathematics, physics, medicine, and other sciences, most statements are not universal. More often than not, our observations about the world involve quantifiers like "most" and "some." There was an important niche for nonuniversal propositions in Aristotle's system of logic.

Ordinarily, if I were to assert, "Some parts of the lecture were interesting," I would most likely be implying that some parts were not interesting. You would certainly not expect me to say that *some* parts were interesting if all parts were. However, the assertion "Some of you will miss a day of work due to illness" does not seem to forgo the possibility that, at some point or another, all of you might miss a day. Oftentimes, in everyday language, "some" means "some but not all," while at other times it means "some or possibly all." To a logician, "some" always means at least one and possibly all.

Some Is Existential

Whereas *all* and *none* are universal quantifiers, *some* is called an *existential* quantifier, because when we use *some* we are prepared to assert that some particular thing or things exist having that description. "Some" propositions are said to be *particular* in nature, rather than universal.

Much like the universal affirmative and negative propositions involving *all* and *none*, Aristotle defined and examined affirmative and negative propositions involving *some*. Whereas the universal affirmative "All people are honest" and the universal negative "No people are honest" cannot both be true, particular affirmations and their negative counterparts are oftentimes both true. The propositions "Some people are honest" and "Some people are not honest" are both most likely true. Medieval scholars named the particular affirmative proposition of the form "Some ***S*** are ***P***" a type-**I** proposition, and they named the particular negation of the form "Some ***S*** are not ***P***" a type-**O** proposition.

With the universal affirmative and negative propositions named **A** and **E**, respectively, students of logic used the mnemonic device—**A**R**I**ST**O**TL**E**—to remember these labels. **A** and **I** propositions were affirmations and come from the Latin ***A****ff****I****rmo* (meaning "I affirm"), and **E** and **O** propositions were negations from *n****E****g****O*** (meaning "I deny").[1] The outer two vowels, **A** and **E** in **A**R**I**ST**O**TL**E**, name the universal propositions, while the inner two, **I** and **O**, name the existential or particular propositions. Medieval scholars also devised a diagram known as the Square of Opposition (Fig. 14) to illustrate the contrary or contradictory relationship between propositions.[2] As seen in the diagram, **I** and **O** are contraries, as are **A** and **E**. For example,

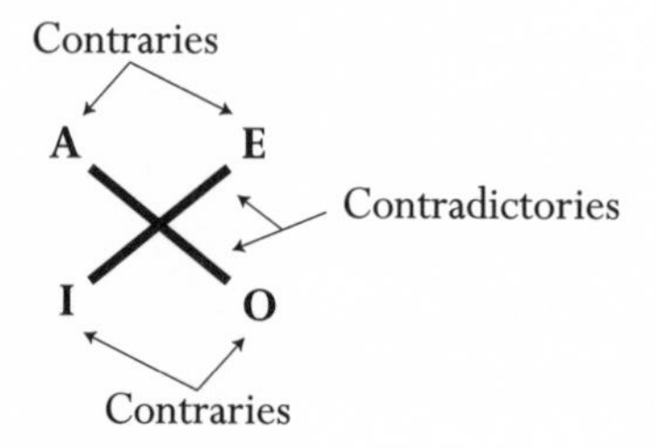

Figure 14. Square of Opposition.

> "Some of you are making noise." "But some of us are not making noise." "Don't be contrary."

The diagonals in the diagram represent the contradictories, **A** with **O,** and **E** with **I**.

Whereas John Venn used overlapping circles for propositions of any type (with different shadings), Gottfried Leibniz, and later Leonhard Euler, used overlapping circles only for expressing particular propositions. To illustrate "Some ***S*** are ***P***," the Leibniz/Euler diagram required the label for ***S*** be written into that part of ***S*** that is in ***P***, whereas for Venn, an asterisk indicated the existence of something in ***S*** that is in ***P***, as shown in Figure 15.[3]

Although the Leibniz/Euler diagram might look a little different if the proposition were "Some ***P*** are ***S***" (the ***P*** would be in

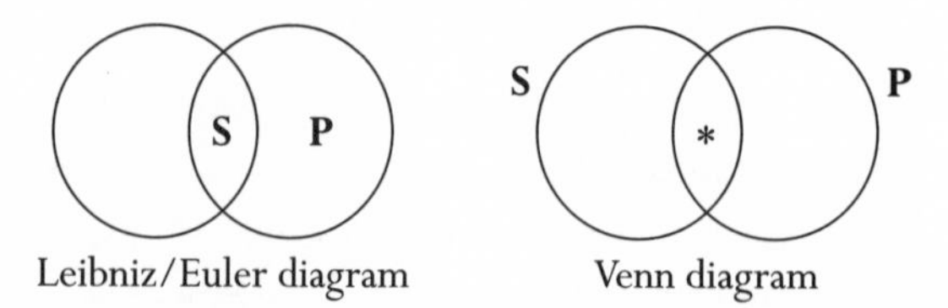

Figure 15. "Some *S* are *P*."

the overlapping region instead of the ***S***), the logicians themselves were well aware that in logic the two propositions are equivalent. Just as "No ***S*** are ***P***" and "No ***P*** are ***S***" are equivalent, "Some ***S*** are ***P***" and "Some ***P*** are ***S***" are interchangeable because their *truth values* are identical. If "Some women are lawyers" is true, then it is also true that "Some lawyers are women." Venn's diagram helps to illustrate this relationship. The asterisk merely indicates that something exists that is both ***S*** and ***P***—as in "Some (one or more) people exist who are both women and lawyers."

However, in our everyday language we do not really use these statements interchangeably. We might hear "Some women are lawyers" in a conversation about possible career choices for women. The statement "Some lawyers are women" might arise more naturally in a conversation about the composition of the population of lawyers. Nonetheless, logic assures us that whenever one statement is true, the other is true, and whenever one is false, the other is false.

Every now and then, "some" statements seem rather peculiar, as in the statements "Some women are mothers" and "Some mothers are women." The first statement is true because some (though not all) women are mothers. The second statement is true because definitely some (and, in fact, all if we restrict ourselves to the discussion of humans) mothers are women. But we must remember that, logically speaking, any "some" statement means "some and possibly all."

For example, we would not normally say "Some poodles are dogs," since we know that all poodles are dogs. During the normal course of conversation, a speaker likes to be as informative as humanly possible. If the universal "all poodles" holds, we generally use it.[4] However, we might say "Some teachers are licensed" if we weren't sure whether all were licensed. Author Jonathan Baron offers the example that when traveling in a new city we might notice that taxicabs are yellow. It would be truthful to say "Some cabs are yellow," withholding our judgment that all are until we know for sure.[5]

Some Are; Some Are Not

An **O** proposition of the "Some are not" form can also be illustrated by two overlapping circles as in Figure 16. Venn's diagram is clearly superior (in fact, the Leibniz/Euler diagram has some serious problems), since "Some ***S*** are not ***P***" and "Some ***P*** are not ***S***" are not interchangeable. Just because the proposition "Some dogs are not poodles" is true does not mean that "Some poodles are not dogs" is. In fact, it is false.

Peter C. Wason and Philip Johnson-Laird have performed studies that seem to indicate that individuals illicitly process "Some X are not Y" to conclude "Therefore, some X are Y," believing they are just two sides of the same coin—in much the same vein as whether the glass is half full or half empty. But in logic the existential quantifier *some* means at least one and possibly all. If it turns out that *all* X are not Y, then "Some X are Y" cannot possibly be true. Their studies indicated that whether an individual gives the material this interpretation depends primarily on the material. Even though subjects were instructed to interpret *some* in its logical fashion, most were able to do so only with material that

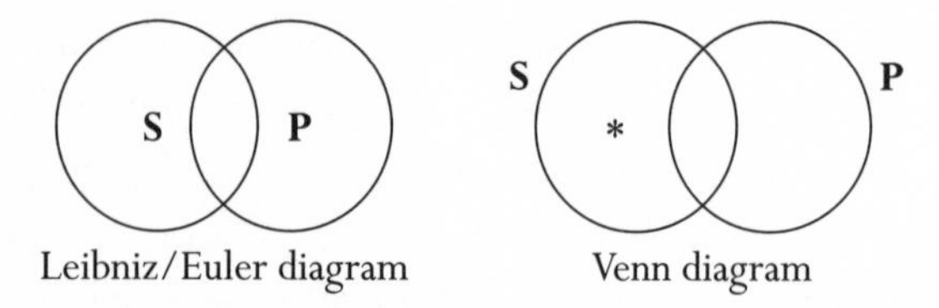

Figure 16. "Some *S* are not *P*."

hinted at possible universality. For example, "Some beasts are animals" was interpreted to mean "Some, and possibly all, beasts are animals," whereas "Some books are novels" was not generally interpreted as "Some, and possibly all, books are novels."

Might this be an indication that we are rational and reasonable after all? A computer could not distinguish between the contexts of these "some" statements in the way that the human subjects did. The subjects in these experiments were reading meaning into the statements given even though they weren't really supposed to. Humans have the unique ability to sometimes interpret what another human meant to say. On the other hand, this tendency to interpret can get us into a good bit of trouble when the interpretation is wrong.

In ordinary language "some" can mean "some particular thing" or "some thing or other from a class of things" and, depending on its use, will signify completely different statements. Compare the statement "Some ice cream flavor is liked by every student" to "Every student likes some ice cream flavor." The first statement indicates a particular flavor exists that is liked by all, while the second statement suggests that each and every student has his or her favorite.[6]

Take a look at Figure 17. Here we have a question taken from the ETS *Tests at a Glance* to introduce prospective teachers to the general knowledge examination required by many states for elementary teacher certification. This question contains examples of many of the concepts we have seen so far. For example, the sentence given is "Some values of x are less than 100" and the examinee is asked to determine which of the answers is NOT consistent with the sentence. The given sentence is a "some" proposition, and the question invokes the notion of consistency with the interference of negation.

The first choice among the answers "5 is not a value of x" is

Some values of *x* are less than 100.

Which of the following is NOT consistent with the sentence above?

A. 5 is not a value of *x*.

B. 95 is a value of *x*.

C. Some values of *x* are greater than 100.

D. All values of *x* are less than 100.

E. No numbers less than 100 are values of *x*.

Figure 17. Sample question from *Tests at a Glance* (ETS). (*Source:* The PRAXIS Series: Professional Assessments for Beginning Teachers, Mathematics (0730) Tests at a Glance at *http://www.ets.org/praxis/taags/prx0730.html.* Reprinted by permission of Educational Testing Service, the copyright owner.)

not inconsistent with the fact that *x* might have some other value that is less than 100. The second choice stipulates "95 is a value of *x*." Indeed, 95 could be a value of *x* since some of the *x*-values are less than 100. The third choice "Some values of *x* are greater than 100" could be true; it is not inconsistent with the fact that some *x*-values are less than 100. Many individuals will probably be tempted to choose choice D as the inconsistent answer, but not if they know that "some" means "some and possibly all." That leaves choice E, which is in direct contradiction to the given statement. If "Some values of *x* are less than 100," then it can't be true that "No numbers less than 100 are values of *x*."

A, E, I, and O

The four types of propositions, **A**, **E**, **I**, and **O**, were the foundation for Aristotle's logic and all that he deemed necessary to develop his rules of logical argument. Aristotle disregarded statements with more than one quantifier—statements like: "Every critic liked some of her films" and "Some critics liked all of her films." Matters could get even more complex if we introduce negation along with more than one quantifier. Consider the following:

Not all of the family enjoyed all of her recipes.

Some of the family did not enjoy all of her recipes.

Some of the family did not enjoy some of her recipes.

All of the family did not enjoy all of her recipes.

By distributing the quantifiers and the negations appropriately, the same basic facts can be articulated in a number of different ways. Although these statements are synonymous, some are easier to grasp than others.[7]

In 1846, Sir William Hamilton of Edinburgh tried to improve on Aristotle's four types of propositions by allowing quantification of the predicate.[8] In his *New Analytic of Logical Forms*, he distinguished eight different forms, defining "some" as "some but only some."

1. All ***A*** is all ***B***.
2. All ***A*** is some ***B***.
3. Some ***A*** is all ***B***.
4. Some ***A*** is some ***B***.
5. Any ***A*** is not any ***B***.
6. Any ***A*** is not some ***B***.
7. Some ***A*** is not any ***B***.
8. Some ***A*** is not some ***B***.[9]

While this system seemed more complete than Aristotle's, there were many difficulties associated with Hamilton's system. His work led to a famous controversy with the English mathematician Augustus De Morgan. One point of disagreement was over Hamilton's definition of *some*. Should "some" mean "some at most" or "some at least" or "some but not the rest"? De Morgan insisted that *some* is vague and should remain so. "Here *some* is a quantity entirely vague in one direction: it is *not-none*; one at least; or more; all, it may be. *Some*, in common life, often means both *not-none* and *not-all*; in logic, only *not-none*."[10] The American logician Charles Sanders Peirce agreed with De Morgan, saying that "some" ought to mean only "more than none."[11]

Hamilton could not really improve upon Aristotle's system; its simplicity had enabled it to remain basically unchanged for two thousand years. With only four types of propositions (**A**, **E**, **I**, and **O**), Aristotle described a structure for logical argument that could be relied upon to yield valid conclusions. His arguments became known as *syllogisms*. Not only would the syllogistic structure always lead to valid conclusions, but as we'll see in Chapter 5, the system could be used to detect rhetoric that led to invalid conclusions.

5

Syllogisms

For a complete *logical argument, we need*
two prim Misses—
And they produce—A delusion.
But what is the whole *argument called?*
A Sillygism.

Lewis Carroll,
Sylvie and Bruno

With the Greek Age of Enlightenment and the rise of democracy, every Greek citizen became a potential politician. By as early as 440 B.C., the Sophists had become the professional educators for those aspiring to a political career and provided them with the requisite instruction for public life. The Sophists were not particularly interested in truth but in intellectual eloquence—some say they were only interested in intellectual anarchy.[1] Plato and later his most famous student, Aristotle, were concerned about those who might be confused by the "arguments" of the Sophists, who used obfuscation and rhetorical ruses to win over an audience. To expose the errors of the Sophists, Aristotle laid down a doctrine for logical argument in his treatise, *Prior Analytics*, or *Concerning Syllogisms*. Indeed, many have said that these laws of inference are Aristotle's greatest and most original achievement.[2]

In *Prior Analytics*, Aristotle investigated the methods by which several propositions could be linked together to produce an entirely new proposition. Two propositions (called the *premises*) would be taken to be true, and another (called the *conclusion*) would follow from the premises, forming a three-line argument, called a *syllogism*. "A syllogism," according to Aristotle, "is discourse in which, certain things being stated, something other than what is stated [a conclusion] follows of necessity from their being so."[3] In other words, a syllogism accepts only those conclusions that are inescapable from the stated premises.

In a syllogism, each proposition is one of Aristotle's four proposition types later classified as types **A**, **E**, **I**, or **O**. The propositions in the first two lines are the premises; the proposition in the third line is the conclusion. If the argument is *valid* and you accept the premises as true, then you must accept the conclusion as true. In his *Letters to a German Princess*, Leonhard Euler said of the syllogistic forms, "The advantage of all these forms, to direct our reasonings, is this, that if the premises are both true, the conclusion, infallibly, is so."[4]

Consider the following syllogism:

All poodles are dogs.
All dogs are animals.
Therefore, all poodles are animals.

The three propositions above form a valid argument (albeit a simplistic and obvious one). Since the conclusion follows of necessity from the two (true) premises, it is inescapable.

Over time, syllogisms were classified as to their *mood*. Since each of the three propositions can be one of four types (an **A** or an **I** or an **E** or an **O**), there are 4 × 4 × 4, or 64, different syllogism moods. The first mood described a syllogism with two universal affirmative premises and a universal affirmative conclusion—named **AAA** for its three type-**A** propositions. The poodle/dog/animal syllogism is an example of a syllogism in mood **AAA**.

A syllogism was further classified as to its *figure*. The figure of a syllogism involved the arrangement of terms within the propositions of the argument. For example, "All dogs are poodles" and "All poodles are dogs" are different arrangements of the terms within a single proposition. In every figure, the terms of the conclusion are designated as the subject and the predicate. If a conclusion reads "All _____ are ______," the term following "All" is called the subject term (***S***) and the term following "are" is called the predicate term (***P***).[5] A conclusion in mood **AAA** reads like "All ***S*** are ***P***." One of the premises includes ***S*** and the other, ***P***, and both include another term common to the two premises, called the middle term (***M***).[6] A syllogism is classified according to its figure depending on the ordering of the terms, ***S***, ***P***, and ***M***, in the two premises. Aristotle recognized three figures, but the noted second century A.D. physician Galen recognized a fourth figure as a separate type.[7] The figures are indicated in Table 1.

Although we could interchange the order of the first and second premises without injury, what we see in Table 1 is the tradi-

Table 1. Syllogism Classifications by Figure

	First Figure	Second Figure	Third Figure	Fourth Figure
First premise	***M-P***	***P-M***	***M-P***	***P-M***
Second premise	***S-M***	***S-M***	***M-S***	***M-S***
Conclusion	***S-P***	***S-P***	***S-P***	***S-P***

tional ordering that was adopted by logicians and brought down to us over the centuries. In fact, psychologists have found that the ordering of the first and second premise can make a difference in how well we perform when reasoning syllogistically. One could even argue that it seems more natural to put the ***S*** in the first premise.

In *Prior Analytics*, Aristotle offered the first systematic treatise on formal logic as an analysis of valid arguments according to their form—the figures and moods—of the syllogism. Historians have noted that in this work Aristotle appears to have been the first to use variables for terms. The idea may have been suggested by the use of letters to name lines in geometry; it is a device that allows a generality that particular examples do not. William and Martha Kneale maintain that this epoch-making device, used for the first time without explanation, appears to be Aristotle's invention.[8] It is not the least bit surprising that the ancient Greeks never developed the use of letters as numerical variables (as we do in algebra) given that it was their practice to use Greek letters to represent numbers.

Aristotle considered only syllogisms of the first figure to be perfect or complete. The first syllogism he discussed was the **AAA** mood in the first figure. The **AAA** mood in the first figure acquired the name *Barbara* in medieval times from the Latin for "foreigners" or "barbarians," with the vowels reminding the scholar or student of the mood—b**A**rb**A**r**A**. In fact, the 14 valid syllogisms identified by Aristotle, along with 5 more added by medieval logicians, were each given mnemonic Latin names to simplify the task of remembering them. When Aristotle explained his first valid syllogism (**AAA**), he generalized the syllogism using Greek letters but for our ease, we'll use the English translation:

All ***B*** are ***A***.

All ***C*** are ***B***.

Therefore, all ***C*** are ***A***.

It is somewhat surprising to the modern mind that Aristotle chose the ordering of the two premises that he did. For example, the first figure **AAA** syllogism might seem more naturally expressed as: All ***C*** are ***B*** (the second premise). All ***B*** are ***A*** (the first premise). Therefore, all ***C*** are ***A***. Or better still, let's keep everything in alphabetical order to maintain the beauty of the transitivity of this argument: All ***A*** are ***B***. All ***B*** are ***C***. Therefore, all ***A*** are ***C***. But what has been passed down to us is: All ***B*** are ***A***. All ***C*** are ***A***. Therefore, all ***C*** are ***A***. This is, perhaps, because the Greek wording does not easily translate to the active voice in the English language. In the Greek, the predicate term appears at the beginning of the sentence and the subject term at the end. The *Student's Oxford Aristotle* translated Aristotle as asserting, "If then it is true that ***A*** belongs to all that to which ***B*** belongs, and that ***B*** belongs to all that to which ***C*** belongs, it is necessary that ***A*** should belong to all that to which ***C*** belongs."[9] Though the wording seems awkward to us, the beautiful transitivity of the syllogism is clearly communicated.

There are three other valid syllogism moods of the first figure—**EAE**, **AII**, and **EIO**. They received the mood-names C**e**l**a**r**e**nt, D**a**r**ii**, and F**e**r**io**, respectively. In our modern day translation, they are:

No ***B*** are ***A***.

All ***C*** are ***B***.

Therefore, no ***C*** are ***A***.

All ***B*** are ***A***.

Some ***C*** are ***B***.

Therefore, some ***C*** are ***A***.

No ***B*** are ***A***.

Some ***C*** are ***B***.

Therefore, some ***C*** are not ***A***.

Notice that these four syllogisms in the first figure produce conclusions of all four types, **A**, **E**, **I**, and **O**.[10]

These are the only four of the 64 moods that produce valid syllogisms in the first figure. Examples of those valid syllogisms—syllogisms "of such form as to be incapable of leading from true premises to a false conclusion"[11] are given in Table 2. In each case, if we accept the two premises as true, the truth of the conclusion is guaranteed to follow. Because there are four different figure arrangements of 64 different moods, there are 4 × 64, or 256, possible syllogisms. Cognitive psychologists have argued that since the order of the two premises can be reversed, there are really 512 different possible syllogisms.[12] Only a few of these are valid, exemplifying correct reasoning.

In the early part of his work on syllogisms, Aristotle had demonstrated how some statements could be *reduced* to other statements.[13] "No pleasure is good" could be translated to "No good thing is pleasurable." "Some pleasure is good" reduced to "Some good is pleasurable."[14] Other propositions, however, were incapable of being reduced; the **O** proposition "Some animal is not man" did not reduce. After offering these specific examples of reductions, Aristotle introduced general reduction rules. Then he set out to prove which syllogisms were valid and which were not and was able to decrease the number of valid

Table 2. Valid Syllogisms in the First Figure

Barbara	All birds are animals. All canaries are birds. Therefore, all canaries are animals.
Celarent	No beans are animals. All chickpeas are beans. Therefore, no chickpeas are animals.
Darii	All biographers are authors. Some curators are biographers. Therefore, some curators are authors.
Ferio	No bases are acids. Some chemicals are bases. Therefore, some chemicals are not acids.

syllogisms to a bare few. By the use of reduction, Aristotle was able to translate most of the valid syllogisms to syllogisms of the first figure while the remaining syllogisms were justified through argument using the law of noncontradiction.

Between the ninth and mid-sixteenth centuries as the English university system evolved, logic or *dialectic* became one of the seven liberal arts. By the latter half of the tenth century, logic had acquired a place of prominence in the curricula at both Cambridge and Oxford universities. As we have seen, scholars devised elaborate methods for classifying the valid syllogisms from among the 256 possible syllogisms. Mnemonic verses provided assistance to students as they were required to memorize the moods and figures of the valid syllogisms. Historians William and Martha Kneale report that the famous mnemonic verses made their first appearance in *Introductiones in Logicam* or *Summulae*, the work of the Englishman William of Shyreswood, in the first half of the thirteenth century.[15]

Barbara celarent darii ferio baralipton
Celantes dabitis fapesmo frisesomorum;
Cesare camestres festino baroco; darapti
Felapton disamis datisi bocardo ferison.[16]

Each word in the Latin verse was a formula wherein the first three vowels indicated the mood (the types of the three propositions) of the valid syllogism. The consonants indicated the rules for reduction. The initial consonant indicated the mood-name of the first figure to which the syllogism was to be reduced. Other consonants provided the steps by which the reduction was achieved.

There were many other tedious rules that logicians brought to the table for the judging of valid syllogisms—rules such as: Every valid syllogism has a universal premise. Every valid syllogism has an affirmative premise. Every valid syllogism with a particular premise has a particular conclusion. Every valid syllogism with a negative premise has a negative conclusion. And so on. Even invalid syllogisms acquired Latin vocables: *Ex mere negativis nihil sequitur; ex mere particularibus nihil sequitur* ("From only negatives, nothing follows; from only particulars, nothing follows").[17] As logician Willard Van Orman Quine pointed out, none of these memory devices and incantations would have been necessary if scholars and students had access to diagrams like Venn's.

Both Gottfried Leibniz in 1686 and Leonhard Euler in 1768 used their circle diagrams to demonstrate the logic behind each of the valid syllogisms. In addition, Leibniz demonstrated each valid syllogism with another diagramming method he had invented—a method using parallel line segments resembling those seen in Figure 18.[18]

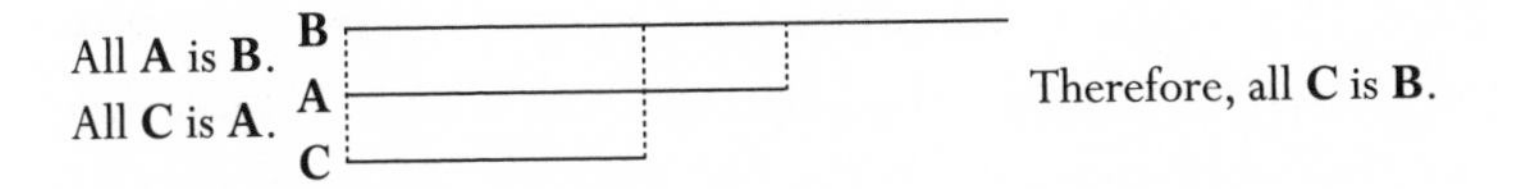

Figure 18. Leibniz's diagram of a *Barbara* syllogism using lines.

Using diagrammatic methods, we can analyze the syllogisms to determine for ourselves which are valid and which are not. There is no need to refer to the medieval taxonomy of rules. Consider the following two syllogisms:

All mammals are warm-blooded animals.

All whales are mammals.

Therefore, all whales are warm-blooded animals.

Some lawyers are Supreme Court justices.

Some women are lawyers.

Therefore, some women are Supreme Court justices.

Both are in the first figure. And in both cases all three statements are true. The issue of validity has nothing to do with whether the conclusion is true but, rather, whether the truth of the conclusion is guaranteed by the truth of the premises. In fact, the first syllogism (**AAA**) is valid (Barbara) and the second (**III**) is invalid. If we check the Latin verse (recalling that the first three vowels of each word gave the mood), we find no mood-name among the mnemonics with vowels **III** regardless of the figure.

When we diagrammed a single proposition, we utilized two circles because two terms (the subject term and the predicate term) were involved. Even though a syllogism has three propositions, it contains only three terms in total. John Venn's diagrams utilized three overlapping circles to analyze a three-line syllogism as seen in Figure 19. The premise "All mammals are warm-blooded" requires the shading of the portion of the mammal circle that is outside of the warm-blooded circle. This portion is shaded to indicate that nothing exists out there. Likewise, for the premise "All whales are mammals," the portion of the whale circle that is outside the mammal circle is shaded to indicate that it is empty. We are left to judge whether our conclusion must necessarily be true. The diagram tells us that it must be the case that all whales are warm-blooded since the only unshaded (non-empty) portion of the whale circle lies entirely inside the warm-blooded circle.

For "some" statements, rather than shading a region (which signifies the absence of something), we need to place an indicator in a region signifying the presence of something. An indicator points out that some things exist that have the quality designated by that region. Venn suggested using numbers to indicate the existence of "some," and we could use different numbers to designate which premise was responsible.[19] To dia-

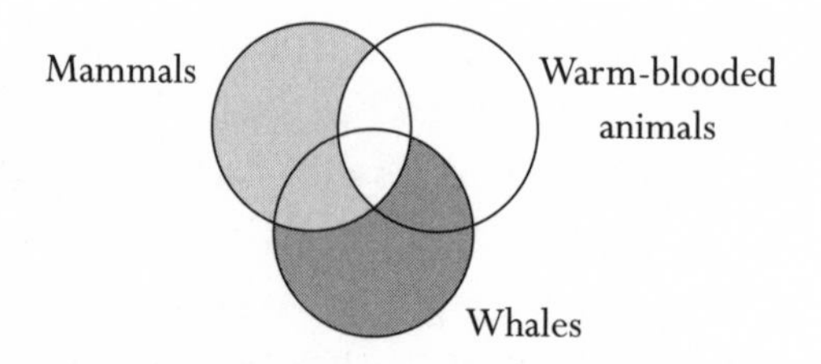

All mammals are warm-blooded animals.
All whales are mammals.
Therefore, all whales are warm-blooded animals.

Figure 19. A Venn diagram for the given syllogism.

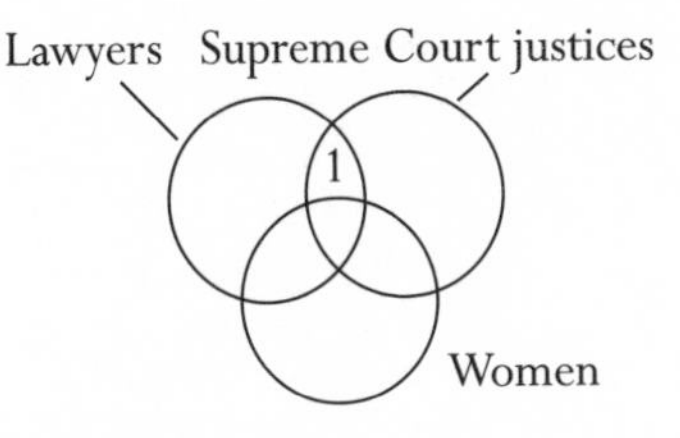

Some lawyers are
Supreme Court justices.

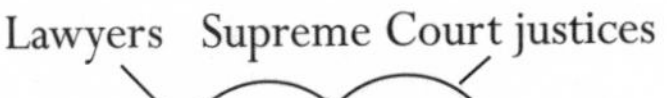

Some lawyers are
Supreme Court justices.

Figure 20. Venn diagrams for the given premise.

gram the premise "Some lawyers are Supreme Court justices," we put a "1" in the overlapping portion of the circles labeled "lawyers" and "Supreme Court justices." The problem is that the overlapping region between lawyers and Supreme Court justices now has two sections—one inside the women circle and one outside it. Our diagram in Figure 20 exhibits one of two possibilities, but we don't know which.

A "2" will designate the regions where "some" possibly exist for the second premise, but we have the same difficulty with the second "some" premise. Since there are two possible diagrams for the second premise as well, for us to examine the inevitability (or lack of it) of our conclusion, we must consult four possible pictures after the two premises are diagrammed. Any one of the four pictures in Figure 21 is a possible scenario based on our two premises. Can we conclude that "Therefore, some women are Supreme Court justices"? One diagram (the first one) indicates that we can conclude nothing of the sort. So, does the conclusion absolutely, undeniably follow from the true premises? No. Substitute "men" for "Supreme Court justices" and the fallacy is revealed by the absurdity of the conclusion.

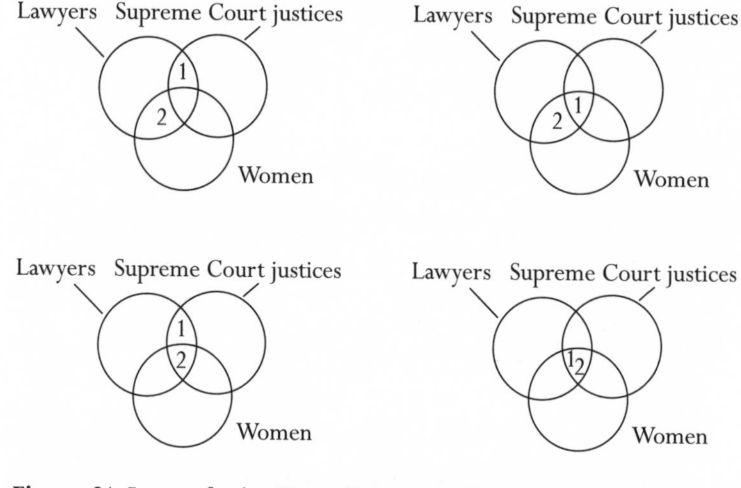

Figure 21. Inconclusive Venn diagrams, all representing "Some lawyers are Supreme Court justices. Some women are lawyers."

Some lawyers are men.

Some women are lawyers.

Therefore, some women are men.

Although it was well known by logicians that syllogisms of the **III** form (like the one above) were not valid, a few logicians of the eighteenth and nineteenth centuries showed how valid conclusions could be drawn if precision was introduced into the two premises. Suppose I say that three (some) of the five pictures hang on the north side of the room, and four (some) of the same five pictures are portraits. Since there are only five pictures in total and three plus four is seven, some must have been double-counted. We can conclude that at least two (some) of the five pictures must be portraits hanging on the north side of the room.[20] This is a syllogism utilizing *numerically definite* quantifiers, rather than vague quantifiers like "some." Although "most" is also vague, a conclusion can also be reached in the above example if "most" means "more than half."

> Most of the pictures hang on the north side of the room.
> Most of the pictures are portraits.
> Therefore, some of the portraits hang on the north side of the room.

Sorites, or Heap

An argument can have more than two premises and more than three terms. The *Oxford English Dictionary* indicates that a *sorites* was an argument composed of a chain of premises in which the predicate term of each premise is the subject term of the next. The conclusion is then formed from the first subject term and the last predicate term. *Sorites* was from the Greek word for "heap" or "pile." In other words, a *sorites* is a heap of propositions chained together to produce one long syllogism, like "All ***A*** is ***B***; all ***B*** is ***C***; all ***C*** is ***D***; all ***D*** is ***E***; therefore all ***A*** is ***E***."

As the number of terms increases, the diagrams used to represent them can get out of hand. John Venn suggested the use of a diagram with four overlapping ellipses such as the one seen in Figure 22 to analyze an argument containing four terms. Each compartment represented a possible combination of truth values for the four propositions. For example, an asterisk is located in a compartment within the ellipses labeled ***A***, ***B***, and ***C*** but outside the ellipse labeled ***D***, so that compartment represents things that are ***A***, ***B***, and ***C*** but not ***D***.[21]

With five terms, ellipses and circles would not do the trick,

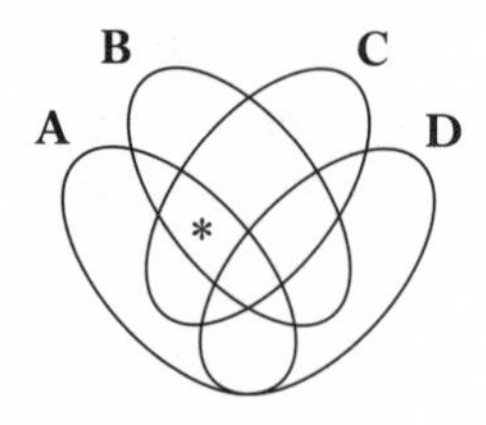

Figure 22. Venn diagram for four terms.

and (after discarding a horseshoe-shaped diagram) Venn proposed the diagram in Figure 23 for a syllogism involving five terms. He also suggested that stamps of the three-, four-, and five-term diagrams could be made to reduce the time required to continually draw them.[22] One glitch in the five-term diagram is the shape of the region representing the class of ***E***-things. Spaces for ***A***, ***B***, ***C***, and ***D*** are ellipses, but ***E*** is shaped like a doughnut with a hole inside. The hole represents a region *outside* the set of ***E***-things.

For diagramming two terms (according to John Venn's methods), we need four separate regions (including the region outside all of the classes). To diagram three terms, we require eight disjoint regions. The diagram for five terms has 31 regions plus the region outside all the enclosed compartments for a total of 32 regions. In general, the diagram for *n* terms requires 2^n regions, or compartments.

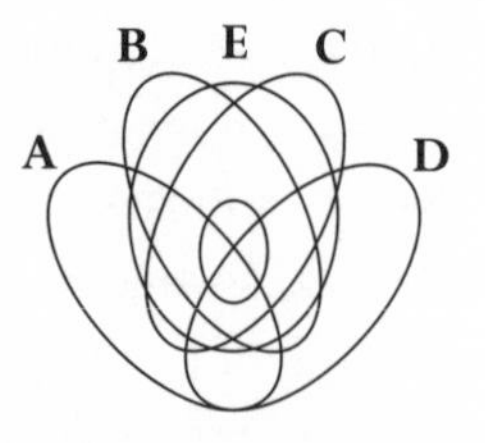

Figure 23. Venn diagram for five terms.

Not long after John Venn proposed his improvements to "Euler" circles, American mathematician Allan Marquand proposed a method of diagrams resembling Mr. Venn's, with two main differences. The shapes of the compartments were rectilinear rather than curvilinear and he assigned a *closed* compartment to the region outside. Marquand demonstrated the analysis of a syllogism with eight terms and indicated that he had had these grid-type diagrams printed up for use cheaply and easily.[23]

In 1886, Marquand's diagram was adopted by Lewis Carroll when he published the solution technique as a game. Regarding the closed compartment representing the outside region, Carroll wrote:

> so that the Class which, under Mr. Venn's liberal sway, has been ranging at will through Infinite Space, is suddenly dismayed to find itself "cabin'd, cribb'd, confined," in a limited Cell like any other Class![24]

Carroll called his entire closed region the *universe of discourse*, a term coined by Augustus De Morgan. Carroll's diagram for five terms with its 32 (triangular) compartments is shown in Figure 24.

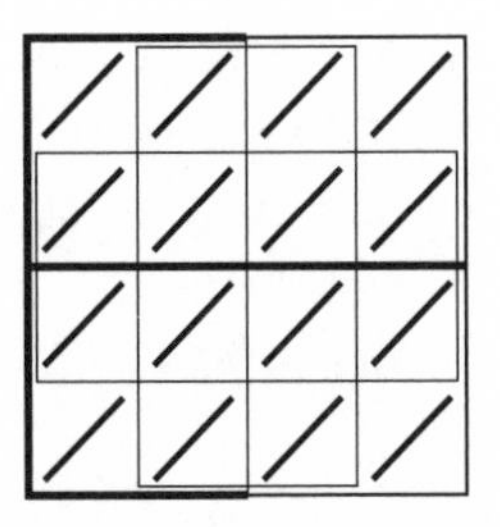

Figure 24. Lewis Carroll's method of diagrams for five terms.

Atmosphere of the "Sillygism"

For all the great advances in diagramming methods, the "human factor" has yet to be overcome. The main difficulty with syllogistic reasoning is that individuals are prone to accept conclusions that do not follow by virtue of logical necessity. A number of attempts by researchers have focused on investigating why syllogisms prove to be so difficult. One early hypothesis put forward by R. S. Woodworth and S. B. Sells in 1935 is called the *atmosphere effect*. They asserted that the moods of the premises created an "atmosphere" regarding the types of conclusions we are prone to accept as valid. For example, individuals are more willing to accept an affirmative conclusion if the premises are affirmations and a negative conclusion if the premises are negations, whether the conclusion follows from the premises or not. In addition, the atmosphere of the premises in terms of whether they are universal or particular predisposes us to accept a conclusion of similar atmosphere.[25] Together, the two premises "Some ***A*** are ***B***" and "No ***A*** are ***C***" create the atmosphere for the adoption of conclusions with a "some" and a negative in them, such as the (fallacious) conclusion "Some **C** are not ***B***."

Perhaps, the effect of atmosphere is not as illogical as it may seem. Upon examination of the valid syllogisms, certain patterns regarding atmosphere were noticed. For centuries scholars followed rules for valid syllogisms like these put forth by Euler, "If either of the premises is negative, the conclusion too must be negative," "If one of the premises is particular, the conclusion too must be particular," and "When both premises are affirmative, the conclusion is so likewise."[26]

The atmosphere of the premises has been shown to be a contributing factor to difficulties in syllogistic deduction, with **I** and

O premises proving to be more difficult than **A** and **E** premises.[27] However, the atmosphere hypothesis alone does not account for other results. Subjects are far more likely to accept erroneous arguments if they can create a "chain" in the argument—like a sorites mentioned earlier. The error of conversion also plays a role in subjects' difficulties with syllogistic reasoning. It is incorrect to think that "All ***A*** are ***B***" means "All ***B*** are ***A***" or that "Some ***A*** are not ***B***" means the same as "Some ***B*** are not ***A***," but individuals often convert these premises when reasoning syllogistically.

Knowledge Interferes with Logic

Oftentimes logical misconceptions such as conversion are more pronounced if everyday or familiar examples are used. This is because individuals invariably try to bring their personal knowledge and experiences to the logical task rather than evaluating the validity of the inference as it stands. For example, suppose I declare truthfully, "All taxicabs are yellow. Your car is not a taxicab." Does it logically follow that your car is not yellow? When examining questions of logic, you must ignore external facts. Don't think about the actual color of your particular car. The correct answer is not yes or no, depending on the paint job of your car. Yet some will answer, "Yes, it follows because my car is green." If I give you less knowledge, you might be more logical. Suppose I declare truthfully, "All taxicabs are yellow. My car is not a taxicab." Is it yellow? Now you can't use knowledge about my car because you haven't seen it. It may be easier to come to the correct conclusion, which is, "Maybe, maybe not."

"All seniors must report to the auditorium. You are not a senior" is likely to be met by the invalid conclusion, "Then I don't have to go to the auditorium." In this case, individuals are likely recalling their past experiences from everyday life. School announcements, such as "All seniors must report to the auditorium," are often made when what is really meant is "All seniors and only seniors must report to the auditorium." We often use language in a manner that is opposed to the requirements of logic. Logic requires that we accept the *minimum* commitments of the premises. In the above example, we have no idea what previous announcements might have been made. "All class officers (they could be sophomores, juniors, and seniors) report to the auditorium" could have preceded the announcement that all seniors report. If we change the example to read, "All members of Mensa are smart. You are not a member of Mensa," it is clear that it does not necessarily follow that you are not smart.

It may be that most of us do not think like logicians. In ordinary language, premises are often inferred. "All dog owners must pay a fee. He owns a poodle." We conclude "Therefore, he must pay the fee."[28] What's missing is the premise "All poodles are dogs." Because this bit of information seems obvious to us, in everyday life we often supply our own premises to form valid conclusions.

Truth Interferes with Logic

"All men are mortals. Socrates is a man." Therefore, what? Given the choices of "Socrates is a Greek" or "Socrates is a mortal" or both, individuals often insert their own factual knowledge into their search for a valid conclusion. We are likely to accept a conclusion that we know is true with little regard for the correctness (or incorrectness) of the inference involved. Although "Socrates is a Greek" may be factually true, it does not logically follow from the premises. Russell Revlis has noted that the difficulty in trying to ignore the facts we know to be true is like the difficulty the juror finds himself in who is asked to disregard inadmissible testimony and base his decision only on the admissible evidence. He emphasized that reasoners do not have insufficient logical skills so much as the inability to distinguish between the information given in the premises and that which is stored in long-term memory.[29]

After a great deal of study, psychologists are still baffled as to the reasons behind our poor performance in syllogistic deduction. Peter C. Wason and Philip Johnson-Laird have investigated the mental processes involved in these deductions and admit this sad state of affairs:

> Syllogistics inference has been studied by psychologists in great detail, yet the process is poorly understood. There is no real theory of deductions, but only a number of scattered hypotheses about the factors which lead to mistakes. One reason for this disappointing state of affairs is the sheer complexity of the quantifiers.[30]

Psychologists have uncovered very little about the ways individuals reason in syllogisms. Perhaps this is not too surprising when we consider all the different arrangements of terms, premises, and moods that can be involved in the seemingly simple construction of a three-line argument.

Terminology Made Simple

The countless scholars devoted to Aristotelian logic established and utilized a new vocabulary—introducing words from the

Latin, such as proposition, universal, particular, affirmation, negation, subject, predicate, premise, and conclusion. The word *logic* had come from the Greek *logos* meaning "reasoning" and *rational* from the Latin *rationation*. Up until the mid-sixteenth century logic was studied exclusively in Latin or the original Greek.

The first known English-language book of logic was printed in 1551, written by Thomas Wilson, and bore the title, *The Rule of Reason, containing the Art of Logic, set forth in English*. Wilson's manual was based on Aristotle's *Organon*, as were most of the works in logic of the time. The book enjoyed considerable success in England for about thirty years and was reprinted five times. Wilson primarily used the Anglicized version of the Latin vocabulary of syllogisms as is our habit and practice today.[31]

In 1573, another Englishman attempted to create an English vernacular for the terminology of logic rather than resort to the modified Latin words. Ralph Lever wrote *The Art of Reason, rightly termed Witcraft* in which he went so far as to suggest we change the word *logic* to *witcraft*. In his search for ordinary English words that could be combined and understood by the commoner, Lever called a "proposition" a "saying" or "shewsay," a "definition" a "saywhat," an "affirmation" a "yeasay," and a "negation" a "naysay." The "premises" were "foresays" and the "conclusion" an "endsay." The subject term of a proposition was the "foreset" and the predicate term was the "backset." So the shewsay, "All dogs are animals," is a yeasay with "dog" as the foreset and "animals" the backset. In an effort to make the terms of logic clear for his countrymen unschooled in Latin, his book contained passages such as the following:

> Gaynsaying shewsays are two shewsays, the one a yeasay and the other a naysay, changing neither foreset, backset nor verb.[32]

He meant:

> Opposing propositions are two propositions, one an affirmation and the other a negation, without changing the subject term, predicate term, or verb.

Would his countrymen have been elucidated by a passage such as this:

> If the backset be said of the foreset, and be neither his saywhat, property, nor difference: then it is an Inbeer. For that we count an Inbeer, which being in a thing, is neither his saywhat, property, kind, nor difference.[33]

Lever's intention of making logic accessible to the masses was admirable and he hoped to make logic less difficult to study and learn by using down-to-earth language. But, thankfully, Lever's colorful "common language" vocabulary did not survive.

Lever was not the only Englishman to try to improve upon the vocabulary of logic.[34] The third Earl Stanhope, Charles, developed some rather peculiar nomenclature for the terms of a syllogism. Stanhope (1753–1816) is known for his many inventions—a microscopic lens, a printing press, an implement for tuning musical instruments, a fireproofing system, a steamboat, and an arithmetical calculator; but what interests us here is his invention of the first instrument for the mechanical solution of problems in logic. Logical proof was called *demonstration* and Stanhope's device was called the "Demonstrator."

Earl Stanhope was anxious to discard the tedious mnemonic verses (Barbara, Ceralent, etc.) that school boys were required to memorize in favor of a simpler system. He showed little respect for the logic curriculum of his time:

> I intend to exclude entirely that long *catalogue of pedantic words* which are now used and which render it generally speaking, both unintelligible to youth and unfit for men of any age. . . . My system of logic will, on the contrary, be found to have the striking advantage of uniting simplicity, perspicuity, utility, and perfect correctness.

Stanhope proposed the Greek word, *holos*, meaning *whole*, as a name for the *middle* term of the syllogism, and *ho* and *los* indicated the subject and predicate terms, respectively. Earl Stanhope's instructions for operating the Demonstrator included this rule: Add *ho* to *los* and subtract *holos*. Stanhope believed his new system of logic had "luminous perspicuity and most beautiful simplicity." Today we may find these attempts at clarifying terminology hilarious.[35]

What had started out as a fairly simple system of valid syllogisms began to acquire a complicated taxonomy with a hideous logical vernacular. The very fact that folks had to memorize moods, figures, or Latin verses to analyze syllogisms might have rankled Aristotle who no doubt thought one could *reason* through arguments. Although we can appreciate the numerous attempts to reframe the vocabulary of logic in simpler language, these efforts do not really simplify the analysis of syllogisms in the way that the diagrams do. Gottfried Leibniz described Aristotle's syllogism as one of the most beautiful inventions of the human spirit. Euler proclaimed it "the only method of discovering unknown truths."[36] "Hence you perceive," he continued, "how, from certain known truths, you attain others before unknown; and that all the reasonings, by which we demonstrate so many truths in geometry, may be reduced to formal syllogisms."[37]

Although these claims may seem exaggerated, the Aristotelian syllogisms, along with the syllogisms of the Stoics, were the bases for all the study of logic that was to follow. However, to follow the syllogisms of the Stoics, we need to introduce an essential word that Aristotle never defined. It is rather surprising that he did not define the word, since he went overboard defining all the other commonly used words. But he simply presumed we understood the meaning of this word, and that word is *if*.

6

When Things Are IFfy

"Contrariwise," continued Tweedledee, "if it was so,
it might be; and if it were so, it would be,
but as it isn't, it ain't. That's logic."

Lewis Carroll,
Through the Looking Glass

Up until now, we've considered propositions of the Aristotelian sort, called simple propositions. At this point, let's consider other, more complicated propositions, formed by connecting simple propositions. The *conditional* proposition, formed by the words *if . . . then . . .*, has been called "the heart of logic."[1] "If" statements come in many forms. Conditionals used to relate two events in time may convey *causality*: "If you press the button, the computer will come on" (the computer came on *because* I pressed the button). "If I get paid on Friday, then I will pay you the money I owe you" is a *promise*. "If you don't do your homework, then you can't watch TV" is a *threat*. A conditional can express an *entailment*, "If the figure is a square, then it has four sides" or present *evidence* for a consequence, "If you earned an A in the course, then you must have worked hard." Frequently the *then* is implied, as in "If wishes were horses, beggars would ride." Reasoning with *if* often proves to be quite tricky.

An ancient story is told by Aulus Gellius about the teacher

Protagoras and his student, a young man named Euathlus.[2] Protagoras agreed to give lessons in oratory and eloquence to Euathlus for a great sum of money, half of which was paid up front and the other half to be paid when Euathlus won his first case in court. Euathlus repeatedly delayed the day of his first court case and eventually Protagoras sued him for the second part of his fee. The master teacher presents his case to the judges, addressing Euathlus and arguing the following:

> If you lose this case, then you must pay me for the judges and the law would have found in my favor. If you win this case, then you must pay me because according to our bargain, you must pay me when you win your first case.

Protagoras must have taught Euathlus well for the young scholar answers:

> If you lose this case, then I owe you nothing by virtue of the decision of the judges and the law. If you win this case, then according to our bargain I shall not pay you because I have not yet won my first case.

The story ends as the judges, reluctant to rule one way or the other, postpone the case.

Deductive reasoning requires a full understanding of the conditional, and whole theories of the word and countless papers have been written on how individuals reason about the word *if*. Conditionals are imbedded everywhere in scientific principles and are essential to our ability to form hypotheses and make logical deductions. Mastery of the logical conditional is crucial to cause-and-effect reasoning, and misunderstandings abound by overinferring as well as underinferring.[3]

Notice that while the Aristotelian propositions connected terms with *all* ____ *are* _____, *no* ____ *are* ____, *some* ____ *are* ____, and *some* ____ *are not* ____, the *if . . . then . . .* connective links entire propositions. In the conditional, "If p then q," p and q represent propositions; p is called the *antecedent* and q is called the *consequent*. For example, in the conditional "If Fifi is a poodle, then Fifi is a dog," the antecedent is the proposition "Fifi is a poodle," and the consequent is the proposition "Fifi is a dog." In a conditional, the consequent necessarily follows from the antecedent. If the antecedent turns out to be true, then we definitely know that the consequent is true. "If p then q" means that whenever p occurs (or is true), q always occurs (or is true). In other words, q necessarily follows from p, or p is said to be sufficient to infer q. Additionally, we can infer that in the absence of q, p could certainly not have occurred. If Fifi is not a dog, then certainly Fifi is not a poodle.

Recall the test discussed in the opening chapter administered by the cognitive psychologists, Peter C. Wason and Philip Johnson-Laird. The subject is shown a blue diamond, a yellow diamond, a blue circle, and a yellow circle (refer back to Figure 1). The examiner announces that he is thinking of a color and a shape; if a symbol has either the color or the shape he is thinking of then he *accepts* it, otherwise he *rejects* it. The examiner accepts the blue diamond. As we analyzed the problem, what we learned from his statement was that one of the other shapes was rejected (but we didn't know which). We can frame the examiner's thinking as a series of *if / then* statements. We know that one of the following conditionals must be true: (1) If he is thinking of blue and diamond, then the yellow circle is rejected; (2) If he is thinking of blue and circle, then the yellow diamond is rejected; or (3) If he is thinking of yellow and diamond, then the blue circle is rejected. Our analysis ended when we realized that because we couldn't determine which "if" was true, we couldn't possibly know which "then" was true.

Interestingly enough, this was the tactic used by mathematician Peter Winkler to confound his opponents with his infamous methods of bidding and signaling in the game of bridge. Called the cryptologic or encrypted methods, they have been declared illegal in tournament play in North America. The bidding system utilizes conditionals, and like the example given above, the opposing team cannot tell from the bidding which of several antecedents in a conditional is true. For example, a bid might mean "If I have the ace and king of the suit you named then I have the ace of clubs and if I have the king and queen of the suit you named then I have the ace of diamonds, but if I have the ace and queen of the suit you named then I have the ace of hearts." Even though everyone at the bridge table is allowed to know that this is what the bid means, only the partner naming the suit (and the one holding the cards) know which "if" holds. Only the partners employing the cryptologic bidding system have access to the information necessary to determine which antecedent is true. Without this knowledge, the opponents can't possibly know what consequent to infer.[4] No wonder it was declared illegal.

Psychologists have found that conditional reasoning with the *if/then* type of statement is extraordinarily difficult. To test higher-order thinking skills such as hypothesis-testing, a relatively simple task was devised by cognitive psychologist Peter C. Wason. First published in 1966, the Wason selection task is said to be one of the most investigated deductive reasoning problems ever constructed. Four cards were shown to the subject and the subject was informed that there was a letter on one side of each card and a number on the other. Four cards such as those shown in Figure 25 were displayed to the subject along with a rule

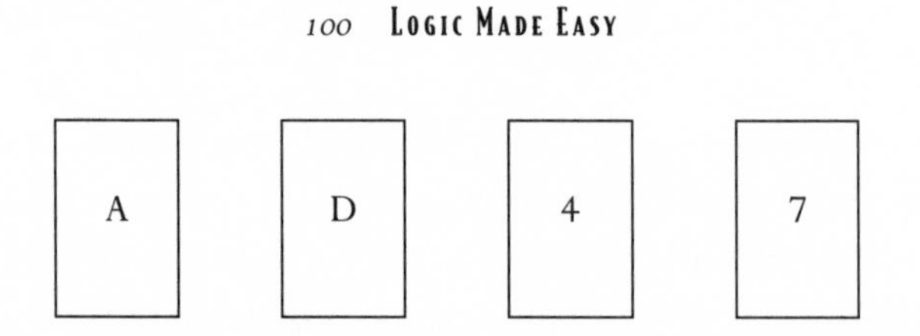

Figure 25. Wason Selection Task. If a card has a vowel on one side then it has an even number on the other.

posed in the conditional:[5] "If a card has a vowel on one side then it has an even number on the other." Furthermore, the rule may or may not be true. The subject was required to name only those cards that needed to be turned over to find out whether the rule was true or false. Subjects rarely selected the correct cards and adults fared as poorly as children. In this study, as well as many others that have replicated it, the correct answer was given less than 10 percent of the time.

The cards showing the **A** and the **7** are the only ones that can lead to discrediting the rule. Subjects see the selection of the **A** card as fairly obvious. If its opposite side reveals an odd number, the rule is discredited. If its opposite side shows an even number, the rule is confirmed. Since the rule says nothing whatever about nonvowel cards, the **D** card is of no interest; we don't care what kind of number is on its other side. And for the same reason, the **4** card is of no interest whatsoever. Even if the **4** card had a nonvowel on the other side, it doesn't invalidate the rule because the rule says nothing about nonvowel cards. However, the **7** card must be selected, for the reason that if the **7** card has a vowel on the other side, the rule is not true.

Subjects usually select the **A** card and the **4** card and sometimes just the **A** card. One interpretation of this mistake is that subjects may think that the rule is actually, "Cards with a vowel on the upper side have an even number on the lower side." The need to apply the rule to the opposite face is not recognized.[6]

However, the way the task is constructed, with some cards number up and others letter up, it seems hard to accept this interpretation of subjects' mistakes.

Regardless of the reasons, in this study and others, subjects showed a clear preference for selecting potentially confirming evidence and neglected potentially falsifying evidence. It seems that our attention is in the wrong place. Due to the formulation of the problem, our attention is focused on the cards named in the rule to the detriment of considering relevant cards that are not named. The mention of values in the rule increases their importance and biases the subject toward choosing them. Referred to as the *matching bias*, people judge as relevant the information named in the rule and yet ignore important alternatives.[7]

There are indications that when exposed to the inconsistencies in their decision making, some subjects were able to overcome their errors. However, many other subjects were unwilling to correct their conclusions even in the face of contradictory evidence. Although subjects were wrong, they were confident that they were right. Even when subjects were shown that turning over the 7 card could falsify the rule (and they acknowledged this), they often dismissed the choice with rationalizations.[8]

The cube task is another fascinating experiment that has been used to assess a subject's ability to reason with *if/then* statements. On each side of a cube is either a square, a triangle, or a circle. The following rule is established: "If one side of a cube has a triangle, then the opposite side has a circle." Subjects are asked to imagine Figure 26. Is it possible that there is a square on the opposite side of the cube? A circle? A triangle? Thirty to 50 percent of the subjects gave the answer that it is possible to have a triangle on the opposite side (the only wrong answer).[9]

Of course, in the English language there are many other ways of conveying an *if/then* conditional. The conditional may also be

Figure 26. "If one side of a cube has a triangle, then the opposite side has a circle."

expressed using words such as *implies*, *never without*, *not unless*, and *only if*.

> If I'm wearing my mittens, then I have my coat on.
> I have my coat on, if I am wearing my mittens.
> Wearing my mittens implies I have my coat on.
> I never wear my mittens without wearing my coat.
> I do not wear my mittens unless I have my coat on.
> I have my mittens on only if I have my coat on.
> Whenever I have my mittens on, I have my coat on.
> Only with my coat on do I wear my mittens.

In logic, these forms are logically equivalent, meaning they always have the same truth value. The conditionals are falsified if and only if "I am wearing my mittens" is true while "I have my coat on" is false. *If* **p** *then* **q** can be expressed as: **p** *never without* **q**; **p** *only if* **q**; **q** *if* **p**; **p** *is a sufficient condition for* **q**; **p** *implies* **q**; **q** *is a necessary condition for* **p**; **q** *is implied by* **p**; or **q** *whenever* **p**. Though they are identical statements in logic, there is no reason to believe that individuals interpret these sentence forms in the same way.

In reasoning and language comprehension, there are several factors to consider. Sentences take on meaning based on the denotative (dictionary) meaning, the linguistic structure (syntax and semantics), and the connotation. Connotation includes the factual and experiential knowledge that we bring to the material at hand. One researcher states, "How people understand and reason with *if . . . then* . . . and *all . . . are* . . . statements is surely very sensitive to the content that fills in the blanks of these statements, that is, to the subject matter being reasoned about."[10] It has been argued that individuals have less difficulty when the material is more relevant and less abstract, and indeed, some studies have reported remarkable improvement in adult performance with meaningful content. Then again, others have reported a failure to perform correctly on tasks with familiar content.[11]

When Peter C. Wason and Diana Shapiro modified Wason's selection task to relate the material to the subject's everyday experiences, performance levels were improved dramatically. This time the subject was told that four cards represented journeys made by the experimenter, each with a destination on one side and a mode of transportation on the other. As in the original Wason selection task, four cards were placed down so that the subject could see only one side of each card. The claim was made, "Every time I go to Manchester I travel by train." The statement may or may not be true. The subject was required to name those cards that needed to be turned over to corroborate or disprove the claim. The task setup is seen in Figure 27.[12]

One issue not addressed in the study was the influence of the different wording of the conditional claim. It's possible that subjects have an easier time evaluating "Every time I go to Manchester I travel by train" rather than "If I go to Manchester, then I travel by train." Wason and Shapiro maintain that it is the familiarity of the material that makes the task easier. When subjects

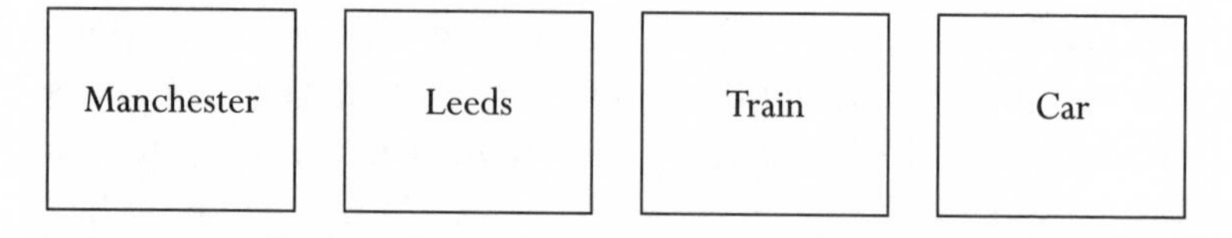

Figure 27. "Every time I go to Manchester I travel by train."

were given a claim to evaluate that made sense to them, 10 out of 16 subjects evaluated the claim correctly (they selected the **Manchester** card and the **car** card). In the control group that performed a similar task by using abstract material and an arbitrary rule (comparable to the original selection task with letters and numbers), only 2 out of 16 performed the task correctly.

In a study conducted by Richard Griggs and James Cox, a task that was logically equivalent to the Wason selection task further demonstrated that subjects do remarkably well when given a familiar rule to evaluate. Florida students were presented with the rule, "If a person is drinking beer, then the person must be over 19." They were then asked to imagine that they were in charge of enforcing this rule. As in the Wason card task, four cards were presented representing persons who may or may not have violated the rule. On one side of each card was the person's age and on the other, the substance the person was drinking. Subjects faced cards labeled "Drinking beer," "Drinking coke," "16 years of age," and "22 years of age." The task was to select those cards and only those cards that were necessary to determine whether the rule was being violated. Seventy-four percent of the subjects were able to choose the correct cards, "Drinking beer" and "16 years of age."[13]

You will recall that in the original Wason selection task, there were two common errors. Most subjects correctly selected the **A** card, most incorrectly selected the **4** card, and most incorrectly neglected the **7** card. This is equivalent to correctly selecting "Drinking beer," incorrectly selecting "22 years of age," and incorrectly neglecting "16 years of age." An individual must search his memory to determine, in this case, "persons over 19" looking for alternatives to "drinking beer" that allow the rule to be true. In other words, an individual must search his memory for an allowable alternative to "drinking beer" for a person of 22 years of age. There are plenty of alternatives that come to mind, consequently individuals don't tend to (mistakenly) select "22 years of age."

The study by Griggs and Cox implemented both concrete materials (person's ages and things they drink) and a familiar rule with a rationale that subjects understood (regarding permission to drink at a certain age). Concrete material alone, however, is insufficient to guarantee consistent reasoning abilities. When subjects were presented with a problem that they had prior experience with, their ability to evaluate a logical rule correctly rose from below 10 percent to over 70 percent. Griggs and Cox concluded that it is not the concreteness of the material being reasoned about that facilitates performance in the logical reasoning tasks but a complex combination of factors. Subjects have difficulty applying a rule of logic when counterexamples in the subject's experience are unavailable or difficult to recall and when the logical task fails to cue individuals to search for counterexamples. When a rule from a subject's past experience is being reasoned about, a falsifying instance from long-term memory can be recalled more easily.

Other studies have replicated the Wason selection task with concrete, but unexplained, rules versus rules in which the subjects were given the rationale behind the rule. One study even went so far as to provide a rule that was counter to common experience: "If a person is over 19 then the person must be drinking beer."[14] When the claim or rule did not conform to the subjects' general knowledge or ran counter to their experiences, the subjects fared no better. However, when subjects knew the rationale behind the rule and it made sense to them, their performance levels showed an impressive improvement.

Arguing that reasoning normally takes place within "linguistic structures" and expressions that encompasses meaning, Herman Staudenmayer takes the position that inferences are not made in isolation but encompass the variety of processing strategies an individual possesses and uses to interpret information.[15] He tested subjects' abilities to evaluate conditional syllogisms with different content and different levels of meaningfulness within the content. He argued that performance in reasoning tasks would be affected by the form of the connective used (*if* ***p*** *then* ***q*** versus ***p*** *causes* ***q***), the use of abstract material (If ***X*** occurs then ***Y*** occurs), and the use of meaningful concrete material versus anomalous concrete material ("If I turn on the switch then the light will go on" versus "If she waters the tropical plant then the light will go on"). In the process of evaluating human reasoning, all of these factors have some effect on an individual's interpretation of the premises and the subsequent evaluation of the conclusion.

Overwhelmingly, subjects misinterpreted the conditional statements. Subjects reasoning with abstract material made more errors than those reasoning with either meaningful or anomalous material. On the other hand, subjects were more consistent with abstract material than with either meaningful or anomalous concrete material. Apparently, individuals had some system of reasoning (even if that system had no relation to the laws of logic) and applied it consistently when reasoning with abstract material where meaning didn't get in the way or jar the senses. In the case of anomalous material, individuals attempt to construct meaning in some hypothetical way (imagining a world where lights are turned on by watering plants). With abstract material, individuals may also attempt to generate a meaningful example about content they understand as a substitute for arbitrary propositions. However, often the example constructed is not one that is applicable under the strict definition of the logical conditional.

Staudenmayer concluded that there are a number of factors that influence a subject's reasoning with conditionals. These include response bias, the number of response alternatives, and the instructions received about the nature of the task. Any theory of logical reasoning needs to distinguish between the subject's interpretation process and the subject's evaluation process in reasoning. Whether an individual accepts a certain interpretation is influenced by general knowledge, presumptions about context, linguistic variables in the sentence, and a predisposed strategy or bias in reasoning. However, the process by which an individual interprets the premises awaits an identification of the precise factors that affect that interpretation.

David O'Brien pointed out that familiarity by itself does not lead to correct inferences.[16] Different domains present different implications and assumptions. The medical and mechanical diagnosis domains, for example, provide a forum for considering logical arguments of the same form that are usually interpreted differently. Suppose a patient is told by a physician that her pain is caused by inflammation and if a particular drug is taken to reduce the inflammation then the pain will go away. We would not necessarily think the physician a liar if the patient did not take the drug and her pain went away nonetheless. Organisms have self-healing properties. Automobiles, on the other hand, are not generally taken to be self-healing. For example, a mechanic tells you that your car is overheating and if the thermostat is replaced the overheating problem will stop. If the thermostat is not replaced and the overheating problem stops anyway, you might be suspicious of the diagnostic abilities of the mechanic.

While familiar conditionals may be easier for us to evaluate, it has been shown that emotional material brings so much baggage to the table that we completely lose sight of the task of evaluating the logic of statements. "If a person convicted of a crime has paid his full debt to society, he should be able to live a private life" might not be evaluated rationally if we find that the person is a child molester.

The Converse of the Conditional

According to Susanna Epp, there is extensive evidence that people perceive "If ***p*** then ***q***" as equivalent to its converse, "If ***q*** then ***p***."[17] This is the identical conversion mistake that individuals make when they think "All ***A*** are ***B***" is the same as "All ***B*** are ***A***." Individuals make this mistake with conditionals hastily and all too frequently, convinced that they are reasoning correctly.[18]

"If the train is traveling from Washington, DC, to Boston, then it stops in New York." I believe this to be a true conditional proposition. If there is a train traveling from DC to Boston, can I conclude that it will stop in New York? (YES.) If I see a train stopped in New York, can I conclude that it is traveling from DC to Boston? (NO.) If I see a train stopped in New York on its way to Boston, can I assume the train originated in Washington, DC? (NO.) If I see that a train originating from DC stopped in New York, can I infer that it is traveling on to Boston? (NO.)

Linguists argue that reasoning difficulties with the conditional are based in our use of language. We very often misuse conditionals. Sometimes, we don't mean what we say; we actually intend the converse of the thought expressed. The statement "If you are over 18, then you are eligible to vote" is the converse of what is meant. Certainly some individuals over 18 are eligible to vote, but it is a fact that all individuals eligible to vote meet the requirement of being over 18. It would be accurate to say, "If you are eligible to vote, then you are over 18."[19]

The medical profession is a field where being able to correctly interpret conditionals is of paramount importance. Yet, in this arena conditional statements are often confused with their converses. When they evaluate medical research, physicians routinely deal with statistics of the sensitivity (true positive) and specificity (true negative) of laboratory test results. In a 1978 study reported in the *New England Journal of Medicine*, it became apparent that physicians often misunderstand the results of these tests.[20] Sometimes interpreted as a difficulty in probabilistic reasoning, it is really a misunderstanding in logic. If a person has the disease, then the probability that they will test positive for the disease is called the sensitivity of the test. If a person does not have the disease, then the probability that they will test negative is called the specificity of the test. For example, the fact that the sensitivity of a test for a certain disease is 0.90 means that the probability the screening test is positive given that the person has the disease is 90 percent. This is called a conditional probability because it is a probability presented in the context of a *logical conditional*. Worded in the *if/then* form, we would say, "If a person has the disease, then the probability of getting a positive test is 90 percent."

Now suppose you take the screening test and get a positive result. What is the probability that you have the disease? No one knows from the information given. We would need different statistics to accurately answer that question. Anyone who claims the answer is 90 percent is confusing the probability of a conditional with the probability of its converse. The probability of having the disease given a positive test result and the probability of receiving a positive test result given that the disease has been contracted are different probabilities that could vary widely. Incredibly, studies indicate that many if not most physicians make this mistake. In a survey of medical literature, David Eddy reported that it's no wonder physicians confused these conditional probabilities; the authors of the medical research often made the error themselves in reporting their results.[21]

People are prone to make the same sort of mistake in evaluating expert witness probabilities in court proceedings. It is not uncommon today to see or hear about DNA experts testifying about the probability that a person who was not the source of the genetic material (the defendant) would nonetheless "match." To say that if the defendant were not the source, the chances are 1 in 20 million that he would match is not the same as saying that if he "matches" the chances are 1 in 20 million that he is not the source. Under certain conditions errors of exactly this sort dramatically increased mock jurors' dispositions to return guilty verdicts.[22]

Investigators who were interested in what aspects of performance led to illicit conversion of conditionals have examined several factors—such as the use of abstract material, the difficulty of the task, the use of binary (yes/no, on/off, etc.) situations, and the negation of the antecedent. Conditionals with negative antecedents have a particular tendency to be interpreted as *biconditionals*.[23] "If you don't see a trash can, then you put the litter in your pocket" is interpreted as meaning "If you don't see a trash can, then you put the litter in your pocket and if you put the litter in your pocket, then you didn't see a trash can." The tendency to (mis)interpret the conditional as the biconditional (the conditional *and* its converse) is universally acknowledged in children and adults.[24]

Several studies have focused on subjects' abilities to reason about the conditional, *if/then*, once alerted to its logical definition. When reasoning about a single conditional, subjects may erroneously believe that *if* **p** *then* **q** also means the converse, *if* **q** *then* **p**, or means the conjunction, *there is* **p** *and* **q**. One study provided evidence that language that articulated the *necessity* of the consequent eliminated error for adults and significantly reduced it for fifth graders. In this study, four reference boxes were prepared, each containing a stuffed animal and a fruit, and a series of conditional statements were made for the subjects to assess as true or false.[25] One fascinating result of the study was the difference in the subjects' interpretations of two false conditionals: the simple conditional, "If there is an apple in the box then there is a horse," versus the conditional where the necessity of the conclusions is made explicit, "If there is an apple in the box then there *has to be* a dog." The experiment is illustrated in Figure 28.

Based on the second statement, when the necessity of the consequent was made clear, fifth graders' correct responses went from 15 to 70 percent, and adults' correct responses went from 75 to 100 percent, a dramatic improvement in both cases.

Other studies have indicated that the conversion and bicondi-

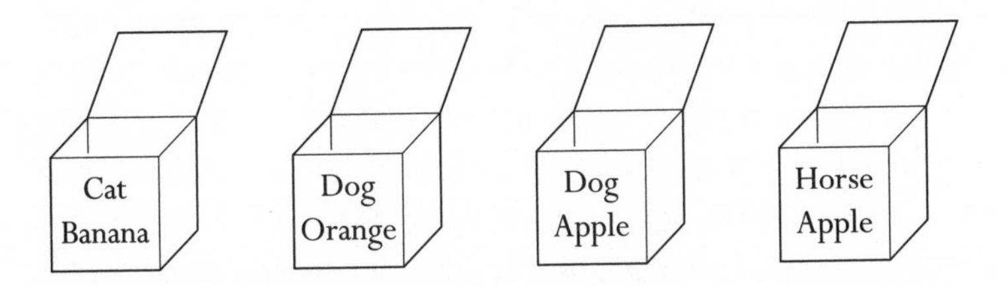

Figure 28. True or False: "If there is an apple in the box then there is a horse" versus "If there is an apple in the box then there has to be a dog."

tional inferences are made less frequently when material is visual or pictorial. You will recall that in Politzer's study, mentioned in Chapter 1, subjects performed much better on a task where information was visual. ("If I wear my dress, I wear my hat." See Figure 4.) In a study by Susan Argent, when subjects were given the statement "If it's a diamond, then it's green," 20 out of 24 (interpreting the statement as a biconditional) considered it appropriate to also infer "All green shapes are diamonds." When given drawings or descriptions of the materials, only 6 out of 24 subjects accepted the mistaken inference.[26] Visual material often aids the reasoner in discovering a counterexample.

There is some evidence that individuals look for a counterexample or contradiction, visual or otherwise, to make correct deductions. Contradiction training in subjects can result in improved performance in deduction skills. In some studies, individuals given information incompatible with their fallacious conclusions showed a tendency to withhold further fallacious inferences.[27]

Causation

Researchers have indicated that the most obvious explanation as to why individuals tend to interpret conditionals as biconditionals is a natural tendency to see the antecedent as *causal*, and furthermore, the *unique cause*, to the consequent.[28] Faced with the conditional "If it rains, then she will spoil her new shoes," individuals may see the rain as the only possible cause of her spoiling her new shoes. Although the rain may indeed be one possible cause, it is not the only one. The temporal nature of the antecedent and the consequent tends to produce a cause-and-effect state of mind.

Peter C. Wason and Philip Johnson-Laird have pointed out that there is a distinct difference in meaning between *if* **p** *then* **q** and **p** *only if* **q** even though they have the same truth value.[29] The forms differ in their connotations of temporal or causal connections. "If the merger is announced, then the stock goes up" and "The merger is announced only if the stock goes up" are logically equivalent. Both conditionals are violated only by the occurrence of the merger announcement without the rise in stock price. However, "If the merger is announced, then the stock goes up" sounds like the announcement causes the stock price to go up or at least the announcement precedes the rise in stock price. On the other hand, "The merger is announced only if the stock goes up" seems to leave vague the causal and temporal connection between the events.

Many conditionals in science require a causal interpretation. A forensic investigator in a drowning case or an ecologist performing a test to determine whether a substance is sea water or fresh may perform a test adding silver nitrate to the substance to test for salt. If the substance is salt, then the reaction will be a white precipitate (it turns cloudy). However, it would be a mistake to think that salt is the unique cause for a substance turning cloudy upon adding silver nitrate. Other substances can produce the same reaction.[30]

Upon encountering *if* **p** *then* **q**, we may infer a causal relation. We may think that **p** is the cause of **q**, that **p** entails **q**, that **q** follows from **p**, or that there is some other correlation between the events, but the proposition itself gives no information whatever regarding the justification for **q** if **p**.[31] Logical conditionals do not *require* a causal interpretation. Individuals' abilities to make strictly deductive inferences are often impaired by misleading causal connotations.[32]

Investigating the difference between pure reasoning (logic)

and practical reasoning, Wason and Johnson-Laird found that the temptation to give temporal or causal interpretations persisted when subjects faced connected materials that invoked practical thought or inference, as well as when subjects faced unrelated materials whose association was abstract. The authors noted, "They are always ready to leave the logical requirements of the task behind and try to establish some meaningful connection between events. . . . The world becomes a more orderly and predictable place if its events are spontaneously organized into a causal matrix."[33] Wason and Johnson-Laird emphasized that it is hardly surprising that causal assumptions take place when drawing practical deductions—hypothesizing causal connections facilitates our ability to draw inferences in the everyday world. They maintained that talk of "making deductions which are valid by virtue of logic alone" and "the distinction between true conclusions and valid inferences" are perplexing ideas that prove to be alien to the ordinary mortal's habitual patterns of thought.[34]

It should surprise no one that we expect causal connections in the conditional when we examine the conditional's historical origins. Historians consider that the analysis of arguments utilizing the conditional and other complex propositions originated with the work of the Stoics. Their idea of what constituted a correct conditional was one in which the notion of the consequent was caused or necessitated by the notion of the antecedent. According to historians William and Martha Kneale,

> When [the Greeks] produced a statement beginning with *If*, they thought of the consequent as being something that followed logically from the antecedent. This is a fairly common use of the conditional form. . . . But it is a mistake to suppose that "if . . . then . . ." is always used in this way, and

much confusion has been produced in logic by the attempt to identify conditional statements with expressions of entailment.[35]

The Contrapositive Conditional

We have seen how it is a mistake to confuse a conditional with its converse. That is because when the conditional statement is true, the converse statement may be true or may be false. "If you exceed the speed limit then you are breaking the law" is true, and its converse "If you are breaking the law then you are exceeding the speed limit" may or may not be true. Another statement, called the *contrapositive*, is always true when the conditional is true (and always false when the conditional is false). The contrapositive of the conditional given above is "If you are not breaking the law, then you are not exceeding the speed limit." Aristotle understood the principle of contraposition when he said, "If it is necessary that ***B*** should be when ***A*** is, it is necessary that ***A*** should not be when ***B*** is not."[36]

The contrapositive of the converse is called the *inverse*. The inverse of the conditional given above is "If you are not exceeding the speed limit, then you are not breaking the law." Like the error of conversion, it is a mistake to believe that its inverse is true just because a conditional is true (mistake of *inversion*). M. Geis and A. M. Zwicky indicate that certain inferences like the inverse are "invited" and the reasoner has a very difficult time not accepting them as valid.[37] As a part of natural language and conversation, *if* ***p*** *then* ***q*** conditional statements that are promises or threats commonly invite the inference, *if not* ***p*** *then not* ***q***. Take, for example, the promise, "If you eat your dinner, you may have dessert." We would probably agree that this promise invites the threat of the inverse, "If you don't eat your dinner, you won't get dessert." But the statement says no such thing. It speaks only of the consequences of eating dinner and says nothing whatsoever about the consequences of not eating dinner. It is curious that even though the common interpretation of this parental warning has no basis in logic, both the parent and the child (and probably all of us) understand the intention of the statement.

I love the example given by Jonathan Baron. Presented with the threat, "If you don't shut up, I'll scream," we would all be surprised if the speaker screamed anyway after you shut up. The speaker probably intends that your interpretation of this conditional include its inverse. "If you don't shut up, I'll scream and if you do shut up, I won't scream." This interpretation may be illogical but it isn't unreasonable; it makes perfect sense. In ordinary discourse, we make practical assumptions about what a person likely means.[38]

Today, a statement such as "If the moon is made of green cheese, then pigs can fly" is considered a true conditional even though both its antecedent and consequent are false. One reason that it is convenient to regard this silly conditional as true is because we would like to consider its contrapositive as true. Its contrapositive, which is just as silly, has both a true antecedent and true consequent. "If pigs can't fly then the moon is not made of green cheese." The reasons for this peculiarity might be clearer with a conditional that isn't nonsense. "If $3 + 1 = 5$, then $3 + 2 = 6$" is true since we would very much like to regard its contrapositive "If $3 + 2 \neq 6$, then $3 + 1 \neq 5$" as true.

If we can avoid the mistakes of inferring the converse, the inverse, or the biconditional, we can make powerful use of the conditional in argument, proof, and scientific hypotheses. Furthermore, all type-**A** universal propositions can be transformed into conditionals. The universal quantification "All ***S*** is ***P***" is easily transformed to the conditional "If a thing is an ***S***, then it is a ***P***." "All mothers are women" becomes "If a person is a mother, then the person is a woman," and "All taxicabs are yellow" becomes "If a car is a taxicab, then the car is yellow." Like Aristotle's categorical syllogisms, conditionals can be used to create syllogisms. As the complexity compounds, we see valid arguments generated using the words *if*, *and*, and *or*. We will next examine those syllogistic arguments.

7

Syllogisms Involving IF, AND, and OR

If the first *and the* second, *then the* third*; but not the* third*; whereas the* first*; therefore not the* second.

Sextus Empiricus

Aristotle's propositions were called *categorical propositions*, since they were constructed from terms or classes representing *categories*. His was a logic of terms and his syllogisms became known as *categorical syllogisms*. As beautiful as they were, Aristotle's syllogisms are not the only forms of syllogism to survive the ages. The Stoics contributed another form of the syllogism, known as the *conditional* or *hypothetical syllogism*, which employed conditional or hypothetical statements. These powerful statements were constructed from entire propositions in lieu of Aristotle's terms. Although Aristotle never addressed conditional statements himself, he used them extensively when establishing the validity of his own syllogisms.

The second century physician Galen is believed to be the author of a tract on logic, *Introduction to Dialectic*, which was discovered in 1844. From Galen's writing, we glean that Aristotle's syllogisms came to be associated with proofs in geometry whereas the Stoics' syllogisms were associated with metaphysical argument. Apparently, the Stoics caused considerable controversy and debate over how the conditional proposition was to be interpreted. According to Sextus Empiricus, the Greek poet Callimachus originated the epigram, "Even the crows on the roof caw about the nature of the conditionals."[1]

The syllogistic logic developed by the Stoics is called *propositional reasoning* as opposed to the categorical or class reasoning of Aristotle. While Aristotle dealt strictly with simple propositions and the ways that terms or classes were assembled together to form these propositions, the Stoics allowed simple propositions themselves to be connected together to form *compound* propositions. One of their connectives was *if*, which generates a conditional proposition. The other connectives they introduced into their logic schema were *or* and *and*. An "or" proposition is called a *disjunctive* proposition, and an "and" proposition is called a *conjunctive* proposition.

Disjunction, an "Or" Statement

The English word *or* can have two different meanings in everyday usage, and we generally rely on context to decipher what the speaker intends. Compare: Coffee or tea? (Not both.) Cream or sugar? (Both are OK.) Was that your husband or your boyfriend? (He can't be both.) Are you coming or going? (You can't do both.) Can you play the guitar or the banjo? (You could play both.) I will get an A in math or history. (I would like to do both.)[2]

Today in logic "or" means "either . . . or . . . or both," but logicians haven't always defined it so. The Stoic logic used what is referred to as the *exclusive* "or," meaning "either . . . or . . . *but not* both." In fact, they most often used *or* when the propositions were diametrically opposed, as in "Either it is day or it is night."[3] As recently as the late nineteenth century, some logicians preferred the exclusive "or" (George Boole, for example). However, modern logic uses *or* in its inclusive sense, as in "You may have sugar or you may have cream (or both)" or "You may send a hard copy or you may send an electronic file (or both)." Sometimes, in ordinary discourse we make use of *and/or* to indicate the inclusive *or*, but in the language of logic, *or* means *and/or*.

Consider the investigation of a problem called THOG. In his study, Peter C. Wason found that the logic of exclusive disjunction proved to be extremely difficult.[4] The subjects were presented with four designs, a black and a white diamond and a black and a white circle, and they were given a rule that defined an invented term, called a THOG. The rule stated that a THOG would have either the particular (unknown) color or the particular (unknown) shape but not both—utilizing the exclusive "or." Given the additional knowledge that the black diamond is a THOG, subjects were asked to determine whether each of the remaining designs was a THOG. Possible answers for each of the three remaining designs were: It must be a THOG, it cannot be a THOG, or it might be a THOG. The design of the task is shown in Figure 29. Researchers found that subjects were likely to be wrong about conclusions involving all three of the remaining shapes. Not only were they likely to be wrong, but the most frequent wrong answers were the exact reverse of the correct solutions because the rule had a built-in element of contradiction.

Subjects frequently declared that the white circle couldn't be a THOG and that the white diamond and the black diamond either might be or must be THOGs. The correct answer is that the white circle is a THOG and the other two are definitely not THOGs. Here is the reasoning: The black diamond is a THOG either by virtue of its black color or its diamond shape but not

Figure 29. The THOG problem.

Rule: In the designs there is a particular shape and a particular color, such that any of the four designs which has one, and only one, of these features is called a THOG.

Given: The black diamond is a THOG. What can you say, if anything, about whether each of the three remaining designs is a THOG?

both. So what makes the black diamond a THOG is either that it is a diamond but not white or that it is black but not a circle. Another THOG would either be not a diamond and white or it would be not black and a circle. Both of these possibilities are satisfied by just one other design—the white circle. Neither of the other two designs can be a THOG. If "diamond" is a THOG feature, then "white" must be the color. The white diamond can't be a THOG because it has both features. The black circle can't be a THOG because it has neither feature. If "circle" is a THOG feature, then "black" must be the color. In that case, the black circle can't be a THOG by virtue of having both attributes and the white diamond can't be a THOG because it has neither attribute. Apparently, even when the definition of the disjunctive is spelled out as clearly as it is in the experiment's instructions, the reasoning can prove extremely tricky.

Conjunction, an "And" Statement

The Stoics defined the conjunctive connective, *and*, in the same way as we define it today. In logic, as well as in our ordinary use of the language, "and" means "both." For example, "You must bring a picture ID and you must answer some questions about your luggage" means both requirements must be met. It is commonplace to abbreviate the propositions themselves. "Phil and Diana are excellent teachers" is merely a shortened form of the statement "Phil is an excellent teacher and Diana is an excellent teacher." The fact that "and" connects two terms instead of two propositions is not a problem since the propositions are implied and the translation is easily made. However, the statement "Phil and Diana make an efficient team" allows no comparable translation and is therefore treated as a single proposition, not a compound one.[5]

"And" statements have been shown to be easier to understand than "or" statements. In fact, "and" statements have been found to be the easiest to grasp followed by "or" statements. Concepts involving both "and" and "or" are the most problematic.[6] In the latter case we must be extremely clear with our use of language. How would you interpret "Sylvester is mean and Spike is lazy or Tweety-bird is smart"? The statement is ambiguous. It could mean that Sylvester is mean ***and*** either Spike is lazy or Tweety-bird is smart. On the other hand, it could mean that ***either*** both Sylvester is mean and Spike is lazy ***or*** Tweety-bird is smart. Sentences like this that take on a different meaning depending on how the sentence is parsed are called *amphibolies*. We should do our best to avoid the amphiboly.

For instance, observe how carefully each of the compound propositions in Figure 30 is worded. The scenario is taken from the GRE Practice General Test. Here, the test-takers are asked questions such as "Which of the following house styles must be on a block that is adjacent to a block that has on it only styles S, T, W, X, and Z?" Choices are Q, R, S, W, and X. Let's reason through this, step by step. Since the given block has an S and an X, the block adjacent to it must have a T and a Z. Any block with a Z on it must have a W. So any block adjacent to the one given

A developer is planning to build a housing complex on an empty tract of land. Exactly seven different styles of houses—Q, R, S, T, W, X, and Z—will be built in the complex. The complex will contain several blocks, and the developer plans to put houses of at least three different styles on each block. The developer will build the complex according to the following rules:

Any block that has style Z on it must also have style W on it.

Any block adjacent to one that has on it both style S and style X must have on it style T and style Z.

No block adjacent to one that has on it both style R and style Z can have on it either style T or style W.

No block can have on it both style S and style Q.

Figure 30. A sample question from the GRE Practice Test. Notice how carefully the "and" and "or" statements are presented. (*Source:* GRE Practice General Test, 1997. Reprinted by permission of Educational Testing Service, the copyright owner.)

must have a T and a Z and a W on it. Since W is the only one of these styles in the answer list, the correct answer must be W.

The founder of the Stoic school of logic was Chrysippus (280–207 B.C.), and it is reported by ancient sources that he and his followers were interested in computing the number of compound propositions that could be constructed from simple propositions by using connectives. From 10 simple propositions, Chrysippus claimed that more than a million conjunctions could be made. Known as the "father of trigonometry," the astronomer Hipparchus of Nicea and Rhodes, who lived during the second half of the second century B.C., said affirmation gave 103,049 conjunctive propositions and negation gave 310,952. It would be interesting to know how they arrived at these numbers. By "conjunction," Chrysippus and Hipparchus probably meant any kind of compound proposition; even so, their numbers defy explanation.

Conditional propositions can always be translated into equivalent disjunctive propositions. In fact, Galen thought that conditionals with negative antecedents were expressed more accurately as disjunctions. Instead of the statement "If it is not day, then it is night," he suggested the "or" statement, "Either it is day or it is night.[7]

Let's see how this works using *or* in the inclusive sense. "If it rains, then I will bring my umbrella" is logically equivalent to "Either it doesn't rain or I bring my umbrella." You will recall that the conditional indicates that if it rains I will definitely bring my umbrella but claims nothing about what I will or will not do if it doesn't rain. In other words, the only time the conditional is false is when it rains and I don't bring my umbrella. Since the disjunctive proposition is true when either one or both of its disjuncts are true, the only time the disjunction, "Either it doesn't rain or I bring my umbrella," is false is when it rains and I don't bring my umbrella. So, "if ***p*** then ***q***" can always be translated to "**not**-***p*** or ***q***." Conditionals also have an equivalent conjunctive form, "**not**(***p*** and **not**-***q***)," which translates to "It is not the case that it rains and I don't bring my umbrella." But we're getting into some hard-to-handle double negatives, so let's just stick with the original conditional form.

Hypothetical Syllogisms

The Stoics advanced Aristotle's theory of syllogisms to include compound propositions, and the word "hypothetical" referred to compound statements, be they conditional, conjunctive, or disjunctive.[8] Chrysippus defined five valid inference schema.[9] The first basic inference of the Stoic schema was: *If the first, then the second; but the first; therefore the second*. This cryptic passage meant:

If the first [is true], then the second [is true].

But the first [is true].

Therefore the second [is true].

Theirs is a three-line syllogism (with two premises and a conclusion) similar to the syllogisms of Aristotle. The Stoics did not use letters as did Aristotle to refer to terms in a proposition, but instead used ordinal numbers, like *first* and *second*. It is not clear whether these words originally referred to terms within a proposition or to the propositions themselves, but the examples they used to illustrate the inference rules employed propositions.[10] Today we take them to refer to propositions. In modern day notation, the Stoics' first inference schema (syllogism) would read like:

If p then q.

p.

Therefore, q.

This is a valid inference as long as the two premises are true. "If you obtain a driver's license in New Jersey, then you must pass a written test. You did obtain a New Jersey driver's license. Therefore, you must have passed the written test." In its first premise, the syllogism contains a conditional proposition with an antecedent and a consequent. The second premise is a simple proposition that affirms the antecedent. This correct inference is called in Latin, *modus ponendo ponens*, or *modus ponens* for short, meaning "mood that affirms." Adults almost never make a mistake on an inference involving modus ponens.[11]

This is exactly the sort of deduction that was used by the earliest attempts at creating machine intelligence. In 1956, artificial intelligence pioneers implemented modus ponens in their program The Logic Theorist, a program designed to make logical conclusions. Given an initial list of premises (true propositions), the program instructed the computer to look through the list for a premise of the form "if *p* then *q*" and a premise *p*. Once these premises were found, the logical consequent *q* was deduced as true and could therefore be added to the list of true premises. By searching for matches in this way, the program used modus ponens to expand its list of true propositions. Armed with modus ponens and some substitution and simplification rules, The Logical Theorist was able to prove an impressive number of mathematical theorems.[12]

Although modus ponens seems like a very simplistic form of deduction, we can use this structure to form elaborate arguments. Consider the following statement: "If you clean up your room and take out the trash, then we can go to the movies and buy popcorn." What do you have a right to expect should you clean up your room and take out the trash? You have a perfect right to expect that we will both go to the movies and buy popcorn. This statement is a conditional of the form: *If **p** and **q**, then **r** and **s*** where "***p*** and ***q***" is the antecedent and "***r*** and ***s***" is the consequent. Utilizing modus ponens, the deduction looks like:

If ***p*** and ***q***, then ***r*** and ***s***.
p and ***q***.
Therefore, ***r*** and ***s***.

Another more elaborate form of modus ponens can be employed by utilizing the law of the excluded middle. One of the premises, the assertion of the antecedent, is often implied.

The following example is familiar to anyone who has completed a United States income tax form:

> If you itemize your deductions, then you enter the amount from Schedule A on line 36.
>
> If you do not itemize your deductions, then you enter your standard deduction on line 36.
>
> Therefore, either you will enter the amount from Schedule A on line 36 or you will enter your standard deduction on line 36.

Symbolically this syllogism is similar to any syllogism of the form:

> If ***p*** then ***q***.
>
> If ***not-p*** then ***r***.
>
> Therefore, ***q*** or ***r***.

The unstated premise is "Either ***p*** or ***not-p***"—the law of the excluded middle—in this case, "Either you itemize your deductions or you do not itemize your deductions." When it is inserted mentally, we know that one or the other of the antecedents is true and therefore one or the other of the consequents must be true.

Some conditionals are relatively easy for individuals to evaluate even when they require the reasoner to envision a large number of scenarios. Most adults would easily negotiate the following: "If your lottery number is 40 or 13 or 52 or 33 or 19, then you win $100." Under some circumstances, we seem to have a singular ability to focus on the pertinent information.

The second valid inference schema of the Stoics was given as: *If the first, then the second; but not the second; therefore not the first*. As the second premise denies the consequent of the conditional premise, this syllogism is known as *modus tollendo tollens*. *Modus tollens* (for short) means "mood that denies" and is considered a more difficult syllogism for most of us to work out.

"If the train is going to Hicksville, then it stops at the Jamaica station. The train did not stop at the Jamaica station." Conclusion? "It can't be the train to Hicksville, because if it were it would have stopped at Jamaica." The symbolic form of the modus tollens argument looks the same regardless of its content. "If ***p*** then ***q***. ***Not-q***." Therefore: "***Not-p***."

Inferences with modus tollens are far more difficult and, not surprisingly, correct responses take longer. It has been suggested that negation makes the inference more difficult. Another theory is that the difficulty occurs because of the direction of the inference (from ***q*** to ***p*** rather than ***p*** to ***q***). In experiments by Martin Braine, difficulties with modus tollens were reversed when the conditional was worded ***p*** *only if* ***q*** rather than *if* ***p*** *then* ***q***. Whatever the reason for the difficulty, modus tollens problems indicate that it is hardly an elementary one-step procedure. The very fact that adults perform well with modus ponens and rather less well with modus tollens suggests that for many modus tollens is not an entrenched pattern of inference.[13]

The inferences of modus ponens and modus tollens are so universal that they appear as two figures of argument in Buddhist logic, called the Method of Agreement and the Method of Difference.[14] The Buddhist system of logic was created in India in the sixth and seventh centuries A.D. under the masters Dignāga and Dharmakīrti. The system had evolved from an earlier five-step syllogism of the school of the Naiyāyiks, which was primarily used for the communication of knowledge to another person rather than discovering knowledge for oneself. The five-step syllogism resembles a mathematical proof in that the first step is one's thesis and the last step, the conclusion, repeats the thesis. Since these five-step syllogisms were used for public argument and explanation, the speaker would want to clearly formulate his thesis from the very beginning. From these five steps, Dignāga's logical reform retained only two steps. Modus ponens, or the Method of Agreement, was a two-line syllogism with a conditional in the first line that included an example as a means to justify the rule. The assertion of the antecedent and the conclusion are combined in the second line:

> Wherever there is smoke, there is fire, as in the kitchen.
>
> Here there is smoke; there must be some fire.

Modus tollens, or the Method of Difference, was formulated similarly:

> Wherever there is no fire, there neither is smoke, as in water.
>
> But here there is smoke; there must be some fire.

The last three syllogisms of the Stoic inference schema contained conjunctions and disjunctions.

> *Not both the first and the second; but the first; therefore not the second.*
>
> *Either the first or the second; but the first; therefore not the second.*
>
> *Either the first or the second, but not the second; therefore the first.*

The logical consequence of negating the conjunction "not both . . . and . . ." is laid out in the first of these syllogisms. The second syllogism defines exclusive disjunction, making us confident that the Stoics were using the exclusive *or*. The last syllogism defines disjunction—for a disjunction to be true, evidently one of the two disjuncts must be true.

Common Fallacies

Two classic fallacies of inference involve the conditional syllogism. "If ***p*** (is true) then ***q*** (is true). ***q*** (is true)." Conclusion? There is none. A common fallacy is to conclude that "***p*** is true." This fallacy leads us to (incorrectly) affirm the antecedent based on the premise that affirms the consequent and entails the error of *conversion* that we have seen time and again. "If there is a stop sign, then you stop the car. You have stopped the car." To conclude "There is a stop sign" would be fallacious. There are many reasons that you might stop the car. The only necessity involved in the *if/then* statement involves what necessarily happens if you encounter a stop sign. The fallacy of affirming the consequent is one of the most frequently made errors in reasoning with conditional syllogisms.[15]

The second fallacy, based on the error of *inversion*, is made by denying the antecedent and leads us to (incorrectly) deny the consequent.[16] Given the two premises "If ***p*** is true then ***q*** is true. ***p*** is not true," it would be fallacious to conclude that "***q*** is not true." For example, "If there is a stop sign, then you stop the car. There is no stop sign." Again, you can infer nothing. The fallacy would be to conclude "Therefore, the car is not stopped." The fallacies of *affirming the consequent* and *denying the antecedent* are named after the second premise ("***q*** is true" and "***p*** is not true," respectively) and not the fallacious conclusion. People of all ages are prone to these fallacies. Modus ponens, modus tollens,

First premise	*if p then q*			
Second premise	*p*	*not-p*	*q*	*not-q*
Conclusion	*q*	*not-q*	*p*	*not-p*
Correct determination	Valid, modus ponens	Fallacy, denying the antecedent	Fallacy, affirming the consequent	Valid, modus tollens

Figure 31. "If/then" syllogistic structures.

affirming the consequent, and denying the antecedent are all instances of syllogisms (the first two valid and the second two invalid) with the conditional premise, *if/then*, followed by one additional simple premise. Figure 31 displays those structures.

One interpretation of why the fallacies of affirming the consequent and denying the antecedent are so prevalent is that subjects interpret the conditional as the *bi*conditional. "If flight 409 is canceled, then the manager cannot arrive in time" is misinterpreted as "If flight 409 is canceled, then the manager cannot arrive in time *and* if the manager did not arrive in time then flight 409 was canceled." Individuals see the antecedent and consequent as being mutually contingent—either both present or both absent. If this theory is correct, it could also explain why young children often have no difficulty with a modus tollens inference. By interpreting the conditional as the biconditional, they should get the modus tollens inference correct but for the wrong reason. Unsophisticated reasoners accept invited inferences everywhere—be they fallacious or valid. However, the theory does not explain adults' difficulty in applying modus tollens.[17]

Several studies have examined whether valid inferences occur more frequently and classical fallacies less frequently when the conditional interpretation is made apparent either explicitly or implicitly.[18] To examine differences in reasoning performance when illicit invited inferences were explicitly blocked, researchers asked subjects to evaluate syllogisms with a simple "if ***p*** then ***q***" versus syllogisms with a more complex but informative premise, "If ***p*** then ***q***, but if ***not-p*** then ***q*** may or may not be true. And if ***q***, then ***p*** may be true or it may be false." It was hypothesized that if the incorrect biconditional interpretation was firmly in place as a reasoning rule then the complex premise would seem contradictory to that interpretation. Subjects were given an additional premise (***p***, ***not-p***, ***q***, or ***not-q***) and asked whether the conclusions ***q*** (modus ponens), ***q*** (denying the antecedent), ***p*** (affirming the consequent), or ***p*** (modus tollens), respectively, were true. The material was abstract (cards with numbers and letters) and the answers allowed were "yes, it must be," "no, it can't be," and "you really can't tell." Subjects had received training in responding "can't tell" so that they would not shy away from that answer category. The results indicated that modus ponens inferences proved to be easy even with the simple conditional premise and the incidence of both fallacies dropped significantly with the more complex conditional.[19] The results clearly indicated that fallacious judgments can be blocked when the distinction between essential and invited inferences is made explicit.

In another experiment, subjects were presented with a simple conditional rule ("If there is a duck in the box, then there is a peach in the box") versus a much more complex rule where additional premises would implicitly block a mistaken line of reasoning ("If there is a pig in the box, then there is an apple in the box. If there is a dog in the box, then there is an orange in the box. If there is a tiger in the box, then there is an orange in the box"). As in the first experiment, subjects were required to evaluate the validity of a conclusion given an additional premise. Once again, most subjects in every age group committed the classic fallacies with the simple premise and most avoided the fallacies with the complex premise. The authors conclude that when presented alone, the conditional "if ***p*** then ***q***" strongly invites the inferences "if ***not-p*** then ***not-q***" and "if ***q*** then ***p***," because these are commonly implied when the conditional is used in conversation, as in promises and threats. We are more cued to the "logic of conversation" than we are to the logical properties of *if*. In laboratory research on reasoning, the conventions of conversational comprehension (and consequently the traditional fallacies) can be set aside by alerting subjects to abandon their interpretations of ordinary discourse. This suggests that fallacious judgments are not explained by faulty reasoning—but rather by language comprehension (our understanding of the definition of "if") and failure to attend to the logical task.

There is an interesting parallel in computer science. The computer language FORTRAN, which stands for FORmula TRANslation system, dates back to 1954.[20] In its early releases, FORTRAN had conditional statements, called IF statements, that were used in its instruction code. In 1980, FORTRAN developers released a new version of the computer language called Fortran 77. This version had, among other features, new ways of handing IF statements. Prior to that time, an IF statement in a FORTRAN program was worded somewhat like a short conditional in an English sentence. "If you turn in your income tax forms on time, you will not be assessed for a late return." In other words, the *then* was missing but implied. With the advent of Fortran 77, the format became IF . . . THEN . . . ELSE. . . . In an English sentence, this would read as follows: "If you turn in your income tax forms on time, then you will not be assessed for a late return, otherwise the IRS will assess late penalties." This statement is certainly clearer and, with this reminder, our attention is drawn to the fact that when the antecedent is true the consequent is definitely true, but when the antecedent is false (the ELSE) then perhaps we need to make the consequences explicit. In any event, the new format for IF statements certainly blocks the computer programmer from making a common mistake.

Diagramming Conditional Syllogisms

Propositional reasoning is defined in terms of the truth or falsity of propositions. The rule of the conditional declares that "if ***p*** then ***q***" is true except when ***p*** is true and ***q*** is false. In modus ponens, we are given that the premise "if ***p*** then ***q***" is true and the premise ***p*** is true; we may conclude that therefore ***q*** is true. We can diagram the premises of this argument by using a Venn diagram and shading any areas we know from the premises to be empty. For example, "if ***p*** then ***q***" tells us that there is no ***p*** without ***q***; consequently any region of the diagram indicating ***p*** without ***q*** would be empty, as shown in Figure 32.

Completing the graph of our premises in modus ponens, we add in the second premise, "There is ***p***." As we did with the "some" statements, we can put a star in the appropriate section, indicating the existence of ***p***. There is only a section of ***p*** that is

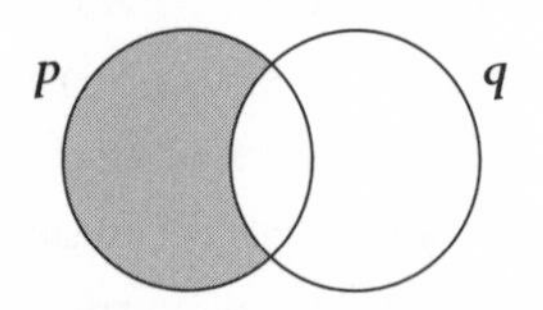

Figure 32. A Venn diagram of "if *p* then *q*."

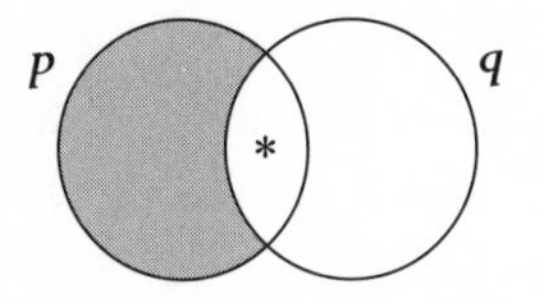

Figure 33. A Venn diagram of "if *p* then *q*. *p*."

left, and we put a star in that section as seen in Figure 33. Conclusion? The star indicates that there is definitely ***q***.

Notice that if the circles were relabeled ***S*** and ***P***, the diagram in Figure 32 would look identical to the type-**A** universal categorical proposition, "All ***S*** is ***P***," where we refer to class inclusion. (All ***S***-things are in the class of ***P***-things.) This makes sense because type-**A** propositions are easily transformed into conditional statements. "All trains that go to Hicksville are trains that stop at Jamaica" can be translated to "If the train is going to Hicksville, then it stops at Jamaica."

Let's examine one of the common fallacies that we discussed earlier. The fallacy of denying the antecedent has as its premises,

If ***p*** then ***q***.
Not-p.

"If ***p*** then ***q***" is diagramed as before—shading the region that is ***p*** without ***q***. For ***not-p***, there are two possible diagrams because there are two distinct possibilities for the location of the star. The star for ***not-p*** could be located outside ***p*** and outside ***q***—sort of hanging outside both of them. Or, the star signifying that there is something that is ***not-p*** could be outside ***p*** but within ***q***. Figure 34 indicates that there are two possible scenarios for the ***not-p*** star. Can we conclude that there is definitely ***not-q***? Not at all. In one case, the star indicates ***not-q***, while in the other case the star indicates ***q***. There could be ***q*** or not, and it would

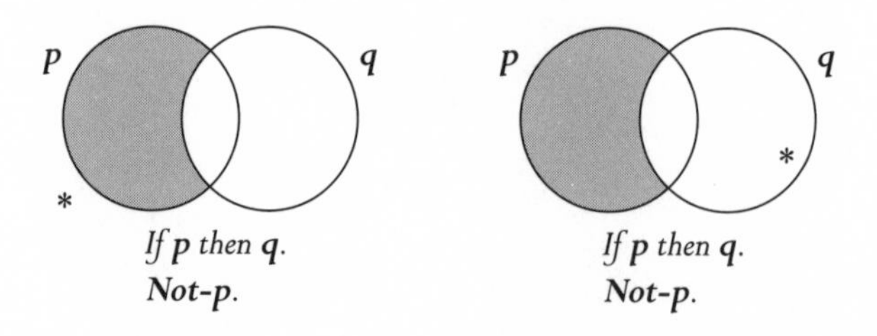

Figure 34. A Venn diagram for possible scenarios for the *not-p* star.

be fallacious for us to conclude otherwise (fallacy of denying the antecedent).

Both categorical logic and propositional logic have seen the introduction of *relational* expressions, which can present all sorts of predicaments. If ***A*** sits beside ***B*** and ***B*** sits beside **C**, does ***A*** sit beside **C**? If the foot bone is connected to the ankle bone and the ankle bone is connected to the shin bone, is the foot bone connected to the shin bone? Relational expressions can be used to create relational syllogisms or series syllogisms. We will explore these in the next chapter.

8

Series Syllogisms

A ham sandwich is better than nothing.
Nothing is better than eternal happiness.
Therefore, a ham sandwich is better than eternal happiness.

UNKNOWN

Of course the ham sandwich syllogism is a joke, but it is an example of a different type of syllogism. A *series syllogism* involves the use of relational phases, such as "is better than," "is older than," "is taller than," "is the mother of," and "is next to." Also called *linear syllogisms*, they elicit a relational inference. Our knowledge of language such as "is taller than" or "is father to" allows us to follow an argument and accept the conclusions that complete a valid inference. The following is an example of such a syllogism; it is valid and rarely confuses us:

Sue is taller than Wendy.
Wendy is taller than Tanisha.
Therefore, Sue is taller than Tanisha.

In the premise, "All ***A*** is ***B***," the relationship between the subject, ***A***, and the predicate, ***B***, is "is." In premises involving a relationship such as "is taller than," we are often required to insert unstated but understood premises. "Wendy is shorter than Sue," although not explicitly stated, is understood to be true any time that "Sue is taller than Wendy" is true, and vice versa. Similarly, we understand that "Isaac is the son of Abraham" can be substituted for "Abraham is the father of Isaac." However, the "is taller than" structure above does not lend itself to a valid syllogism with the relation "is the father of."

Abraham is the father of Isaac.

Isaac is the father of Jacob.

Therefore, Abraham is the father of Jacob. (!)

These types of syllogisms may have first been mentioned by the physician Galen in his work, *Introduction to Dialectic*. Galen presented "relational syllogisms," which he said fit neither the Aristotelian nor the Stoic logic. "Theon has twice as much as Dio, and Philo twice as much as Theon; therefore Philo has four times as much as Dio." Galen also offered the example, "Sophroniscus is father to Socrates; therefore Socrates is son to Sophroniscus."[1]

Thomas Hobbes, the British author of natural and political philosophy, began his 1655 work, *Elements of Philosophy Concerning Body*, with a section entitled "Computation or Logic." In this exposition, Hobbes offered the example of a sophism wherein the deception lies hidden in the form of the syllogism. His example also reveals the traps inherent in some types of relational propositions: *The hand toucheth the pen. The pen toucheth the paper. Therefore, the hand toucheth the paper.*[2] One reason that this is such a good example of fallacious reasoning is that the reader immediately forms a mental picture from the two premises. Having formed that image, we have to smile when we reach the preposterous conclusion. Hobbes explained that there are really four terms rather than the apparent three terms: *hand*, *touching the pen*, *pen*, and *touching the paper*. The terms *touching the pen* and *pen* are not equivalent.

The hand is *touching the pen*.

The pen is *touching the paper*.

Therefore, *the hand* is *touching the paper*.

Augustus De Morgan, once called the father of the logic of relatives,[3] included the following puzzle from an old riddle book in his discussions:

> An abbess observed that an elderly nun was often visited by a young gentleman, and asked what relation he was. "A very near relation," answered the nun; "his mother was my mother's only child" which answer, as was intended, satisfies that abbess that the visitor must be within the unprohibited degrees, without giving precise information.[4]

Can you figure out the relation of the young gentleman to the nun? Perhaps the elderly nun was answering in such an elliptical fashion because she didn't want the nosy abbess to know that the young man was her son.

Building on the work of De Morgan, Charles Sanders Peirce contributed much to the study of logical relatives, including these examples of relatives in his body of work: ***A*** is given in marriage to ***B*** by ***C***; ***A*** praises ***B*** to ***C***; ***A*** maligns ***B*** to ***C***; and ***A*** praises everybody to somebody whom everybody maligns to ***B***.[5] Peirce then applied a sort of algebra to the study of the logic of relations where he defined the union (disjunction), intersection (conjunction), relative product, relative sum, complement (negation), and converse of relatives.

Should we find ourselves in the position of believing that things are getting far removed from arguments we will ever have to face or evaluate, let's examine a national standardized test. In 1977, the Graduate Record Examination (GRE) General Test was altered by the addition of an analytical measure.[6]

The new measure was introduced to include more aspects of reasoning than had been previously included in the verbal and quantitative measures alone and consisted of test items in logical reasoning and analytical reasoning. Logical reasoning test items assess critical reasoning skills by testing one's ability to recognize assumptions, analyze evidence, and evaluate arguments and counterarguments. Analytical reasoning items primarily involve deductive thinking; they assess the ability to deduce information from a given structure of relationships. Figures 35 and 36 display such questions taken from the GRE test administered by the Educational Testing Service (ETS).

In Figure 35, the relational expression "is next to" or "is adjacent to" is to be analyzed, and the ordering of the cups involves another relation. Try the problem for yourself; then read on. If the magenta ball is under cup 1, then the red ball must be under cup 2 (it is immediately adjacent to magenta). The green ball is under cup 5. So we know the order of the six balls must be MR_ _G_. Now purple must be under a lower number than orange, so purple has to be in the third or fourth position, either MRP_G_ or MR_PG_. Orange has to be under a higher-numbered cup, so our final possibilities are MRPOGY or MRPYGO or MRYPGO. Now check the answers. Answers B, C, D, and E offer pairs of colors that *could* be adjacent but aren't of necessity adjacent. The correct answer is A. Green and orange are adjacent in every scenario.

In Figure 36 the relations "in the same year," "in the previous year," and "in the next year" must be assessed. The ETS claims that questions of this type measure the reasoning skills developed in almost all fields of study and that no formal training in logic is necessary to do well on these questions. How good are your logical (analytical) abilities? See if you can answer the question in Figure 36. We know BCK (beans, corn, and kale) are

In a game, exactly six inverted cups stand side by side in a straight line, and each has exactly one ball hidden under it. The cups are numbered consecutively 1 through 6. Each of the balls is painted a single solid color. The colors of the balls are green, magenta, orange, purple, red, and yellow. The balls have been hidden under the cups in a manner that conforms to the following conditions:

> The purple ball must be hidden under a lower-numbered cup than the orange ball.
>
> The red ball must be hidden under a cup immediately adjacent to the cup under which the magenta ball is hidden.
>
> The green ball must be hidden under cup 5.

Question: If the magenta ball is under cup 1, balls of which of the following colors must be under cups immediately adjacent to each other?

A. Green and orange
B. Green and yellow
C. Purple and red
D. Purple and yellow
E. Red and yellow

Figure 35. Sample analytical reasoning question with relation "is adjacent to." (*Source*: GRE Practice General Test, 1997. Reprinted by permission of Educational Testing Service, the copyright owner.)

planted in the first year. Since he never plants kale two years in a row, he can't repeat K in the second year. Since he always plants beans when he plants corn, he can't plant corn in the next year because he plants no more than one of the same vegetables in the next year. The farmer plants beans, peas, and squash in the second year. Using parentheses to indicate the different years, the

A farmer plants only five different kinds of vegetables—beans, corn, kale, peas, and squash. Every year the farmer plants exactly three kinds of vegetables according to the following restrictions:

If the farmer plants corn, the farmer also plants beans that year.

If the farmer plants kale one year, the farmer does not plant it the next year.

In any year, the farmer plants no more than one of the vegetables the farmer planted in the previous year.

Question: If the farmer plants beans, corn, and kale in the first year, which of the following combinations must be planted in the third year?

A. Beans, corn, and kale
B. Beans, corn, and peas
C. Beans, kale, and peas
D. Beans, peas, and squash
E. Kale, peas, and squash

Figure 36. Sample analytical reasoning question with relations "next" and "previous." (*Source*: GRE Sample Test Question, 1996. Reprinted by permission of Educational Testing Service, the copyright owner.)

first two years must look like: (BCK)(BPS). From the answers for planting in the third year, we can eliminate answer E also because he can't repeat both peas and squash. All of the other answers include beans. If he plants beans again, he can plant neither peas nor squash. This eliminates the answers B, C, and D. This leaves only answer A—beans, corn, and kale. Since A doesn't violate any of the restrictions, it is the combination that must be planted.

Creating mental pictures and spatial images can assist us in reasoning logically. Undoubtedly, most of us create a mental image to reach the conclusion in the following syllogism:

The black ball is directly beyond the cue ball.

The green ball is on the right of the cue ball, and there is a red ball between them.

Therefore, if I move so that the red ball is between me and the black ball, then the cue ball is on my left.[7]

Sometimes, however, mental imagery is not enough. The GRE test analytical reasoning questions are most likely easier if the reasoner's imagination is assisted by diagrams or pictures. At the youngest ages, reasoning is easier with concrete objects or pictures of the objects. Adults don't actually need to draw balls or pictures of vegetables, but constructing some iconic representation of the hierarchical arrangement in the problems, such as an arrangement of letters, can prove to be of invaluable assistance when conceptualizing the possible combinations.

Relational premises can involve *spatial* inclusion, as in "The woman is in the room. The room is in the house. The house is in the town." Also included are *temporal* relations, like "The dinner is in the evening. The evening is in September. September is in the fall." These relations are asymmetrical. The woman is in the house; few, if any, of us with a basic understanding of language would make the mistake of saying that the house is in the woman.[8]

One difficulty inherent in the syllogisms of relations is associated with the ordering of the items in the premises and the ordering of the premises. Consider the three different wordings in the following seriation problem:[9]

1. If stick **A** is longer than stick **B** and shorter than stick **C**, which stick is the shortest?
2. If stick **A** is longer than stick **B** and stick **C** is longer than stick **A**, which stick is the shortest?
3. If stick **B** is shorter than stick **A** which is shorter than stick **C**, which stick is the shortest?

For most individuals the wording in example 3 is the easiest—**B** shorter than **A**, and **A** shorter than **C**—where the object of the first premise is the subject of the second. We conclude, therefore **B** is shorter than **C** and **B** is the shortest. The principle of *end-anchoring* influences the ease of making this conclusion; it is easier to understand a premise whose grammatical subject, rather than its grammatical object, is the subject or predicate of the conclusion.[10]

Example 2 is somewhat more difficult because the order of the premises is not "natural" and must be reversed in our minds. Once we reverse the order of the two premises—**C** is longer than **A** and **A** is longer than **B**—then the first premise is end-anchored with **C**, the grammatical subject of the first premise and the logical subject of the conclusion. Example 1 is the most difficult since either "**A** is longer than **B**" must be translated to "**B** is shorter than **A**" or "**A** is shorter than **C**" translated to "**C** is longer than **A**" for comparison.

Psychologists Peter C. Wason and Philip Johnson-Laird conclude that syllogisms vary widely in their degree of difficulty. Some are straightforward and can be solved in seconds; others are extremely difficult, taking considerable time to solve. Still, even the easiest categorical or propositional syllogisms tend to be more difficult than the three-term series problem.[11]

Long before there was a need to translate language into symbols to input instructions into a computer, some visionaries foresaw the advantages of being able to translate propositions, relational expressions, and even syllogisms into symbols. Once this idea was received favorably, the mathematics of arithmetic could be used to dissect logical arguments.

9

SYMBOLS THAT EXPRESS OUR THOUGHTS

Now the characters which express all our thoughts will constitute a new language which can be written and spoken.

GOTTFRIED LEIBNIZ

Often described as a universal genius, Gottfried Wilhelm Leibniz (1646–1716) excelled at whatever he applied himself to, and his range of expertise in countless fields is astounding. Leibniz contributed to many diverse disciplines, including mathematics, law, religion, politics, metaphysics, literature, history, and logic. He imagined a world in which all thought could be reduced to exact reasoning and all reasoning to exact computation.[1] When he was just 14 years old, Leibniz advocated the utilization of symbols or pictures to simplify complicated logical arguments and called his idea of a symbolic language *lingua characteristica universalis*, or *universally characteristic language*. Through the use of a universal language, individuals and countries could settle feuds as easily as settling a dispute over the sum of a column of numbers. Leibniz optimistically dreamed, "When controversies arise, there will be no more necessity of disputation between two philosophers than between two accountants. Nothing will be needed but that they should take pen in hand, sit down with their counting tables and (having summoned a friend, if they like) say to one another: Let us calculate."[2]

Leibniz wanted to construct an alphabet of human thought—a universal method of expression that anyone of intelligence could read or speak no matter what his native language. In 1677, he stated:

> Those who write in this language will not make mistakes provided they avoid the errors of calculation, barbarisms, solecisms, and other errors of grammar and construction. In addition, this language will possess the wonderful property of silencing ignorant people.[3]

Perhaps his suggestion that a superior form of communication could actually silence ignorant people is a little fantastic, but Leibniz, ever the idealist, never let the complexity of a problem interfere with his goals.

Leibniz proposed the replacement of each term in a proposition with its "symbolic number" and eventually the replacement of the numbers with other signs, possibly symbols that resemble geometrical shapes or Chinese characters. As a transition measure, numbers could be used to represent concepts since they are certain and determinate; furthermore, more complex concepts could be expressed as products of numbers. Leibniz gave the example, let 2 = rational and 3 = animal, then 6 = man, because man = rational animal = $2 \times 3 = 6$.[4] With such a system, he suggested, we could then determine the truth of statements. In the proposition "All men are animals," we examine whether "man" is divisible by "animal." If the symbolic number for "man" (6) is divisible by the symbolic number for "animal" (3), then the statement is true. Leibniz called "rational animal" a composite concept, as it is a composite of more basic concepts and the product of their symbolic numbers. This is perhaps the origin of the modern use of *composite number* (as opposed to *prime number*) to mean any number that can be expressed as the product of whole numbers greater than 1.

In one of his many, many treatises on the universal language, Leibniz asserted, "The whole of such a writing will be made of geometrical figures, as it were, and of a kind of picture—just as the ancient Egyptians did, and the Chinese do today."[5] Through correspondence with a friend working in China, Leibniz had apparently become enamored of the possibility of combining the best of the East and the West. Although the germ of the idea of constructing an artificial language can be seen in the writings other mathematical giants such as René Descartes and Blaise Pascal who preceded him, Leibniz's passion and vision for this idea was unequaled.

Leibniz's dream of a universal language was largely ignored, however; his scientific and mathematical contemporaries must have considered his goal of a universal language an idealistic idiosyncrasy. For one thing, scholars of that time already had an international language—Latin. However, by about the middle of the nineteenth century, Latin began to fade as the language of scholars. In an attempt to separate themselves from Church Latin, to educate even those not competent in Latin and Greek, and to advance growing nationalism, scientists began to publish their works in their native languages. Beginning in about 1880 and continuing until the beginning of World War I, serious efforts were made by a few esteemed scientists of the day to construct an easily acquirable international language. *Volapük*, *Esperanto* (meaning "one who hopes"), *Idiom Neutral*, *Interlingua*,[6] and *Ido* (meaning "offspring" in Esperanto) were all efforts to create the perfect international language. One of the prime movers in this effort was the French philosopher and mathematician Louis Couturat.[7] Couturat is of particular interest to us because he was the individual who made the landmark contribution of bringing to light Leibniz's previously unpublished work on logic. In 1903, Couturat published *La Logique de Leibniz*, over two hundred years after it was written.

Perhaps influenced by the Chinese *Yi Ching*, or *Book of Changes*, and its 64 hexagrams composed of the symbols of yin and yang, Leibniz developed a system of binary arithmetic centuries ahead of his time. The yin and the yang divide the universe into dualities—male/female, yes/no, and on/off. In 1679, Leibniz invented a system of binary numbers and binary arithmetic.[8] He showed how any of our numbers could be represented using only 0s and 1s and demonstrated multiplication and addition in his binary system. Incredibly, binary notation is used in exactly this way in a modern digital computer. A *bit*, the smallest unit of data in a computer, stands for *binary unit* and has a single binary value, either 0 or 1. *Bytes* are made up of bits, usually eight of them. If you have a file on your computer that is 28 KB in size, the file contains 28 kilobytes, or 28,000 bytes of data.

Leibniz also invented a mechanical device for computing, although it did not make use of binary arithmetic. In January 1673 he demonstrated his calculating machine to the Royal Society of London, and the following April Leibniz was elected as a fellow of the Royal Society.[9] Leibniz said that he had come up with the idea for his device when he first saw a step counter, an instrument for recording the number of steps taken by a pedestrian. Developing a similar type of apparatus, he created a calculating machine called a stepped drum calculator or stepped wheel calculator. Leibniz's device could add, subtract, multiply, and divide; it is said that once perfected it would even take square roots! Most of us probably first encountered a four-function calculator with a square root key in the late 1960s, very nearly three hundred years later.

Leibniz's contributions are primarily his original ideas. His idea of using diagrams to analyze syllogisms, his idea of a symbolic language of reason using characters that might be manipulated like an algebraic equation, his idea of using binary arithmetic, and his idea of creating a labor-saving device for calculation were all precursors of what was to come hundreds of years later. From the mid-nineteenth century through the turn of the twentieth century, there was a tremendous push to reduce logic to a powerful, yet simplistic form similar to the algebraic systems dominating mathematics. Mathematical historian E. T. Bell maintains that Gottfried Leibniz and Augustus De Morgan dreamed of adding logic to the domain of algebra, but George Boole was the one who actually did it.

Beginning in 1827 and for forty years thereafter, Augustus De Morgan influenced a number of budding mathematicians in his position as mathematics professor at the University School, London. His main contribution to the development of logic was in reviving interest in the subject through his writing and teaching and through his quarter-century feud with Sir William Hamilton, Professor of Logic and Metaphysics at Edinburgh, Scotland. The quarrel began when Hamilton accused De Morgan of breaking a confidence and plagiarizing his results. In a series of pamphlets and articles with each man viciously attacking the other, the controversy was both unseemly and embarrassing, but De Morgan seemed to enjoy it. It is said that De Morgan valued his enemy, being that the dispute did much to bring De Morgan's name and ideas to the attention of other logicians.[10]

De Morgan introduced the idea of *negative terms* and the notion of the *universe of discourse* to modern logic. His negative terms (De Morgan called them *contraries*, not to be confused with Aristotle's propositions called *contraries*) were symbolized by lowercase letters. If *X* stands for the term *man*, then *x* stands for *not-man* (today we call this the *complement*), and together they

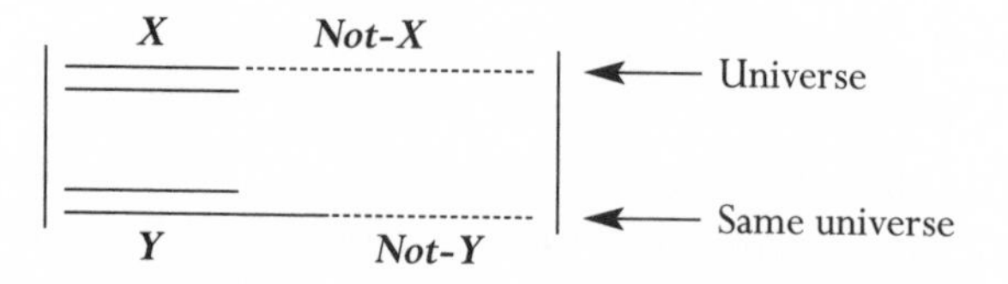

Figure 37. De Morgan's graphical representation of "Every *X* is *Y*." ***The two solid Y-lines illustrate the two cases.***

exhaust the universe of discourse. In other words, everything under discussion must be either "man" or "not-man" (law of the excluded middle).

In an 1850 paper on logical syllogisms, De Morgan mentioned the graphical representations adopted by Euler, but like everyone else, De Morgan appeared to be unaware of Leibniz's work in logic.[11] De Morgan introduced a diagrammatic method that he felt was superior for illustrating propositions. Curiously, his diagrams bear an amazing resemblance to the line diagrams originating from Leibniz's 1686 paper. De Morgan demonstrated "All ***X*** are ***Y***" or "Every ***X*** is ***Y***" as shown in Figure 37.

The solid part of the ***X*** line segment represented things that are ***X*** things, while the dotted part of that line segment represented things that are ***not-X***. The same held for ***Y***, but here De Morgan has illustrated the two possible cases that are the cause of much confusion: ***X***-things could constitute *all* of the ***Y***s (hence the solid segment that is the same length as the ***X*** segment) or only some of the ***Y***s (the longer solid line segment). De Morgan went on to provide an example of a more complicated proposition involving a disjunction, "Some things are neither ***X***s nor ***Y***s" as is seen in Figure 38.

Without a doubt, De Morgan thought that the laws of mathematics could be brought to bear in logic. He says, "Logic considers the *laws of action* of thought: mathematics applies these *laws*

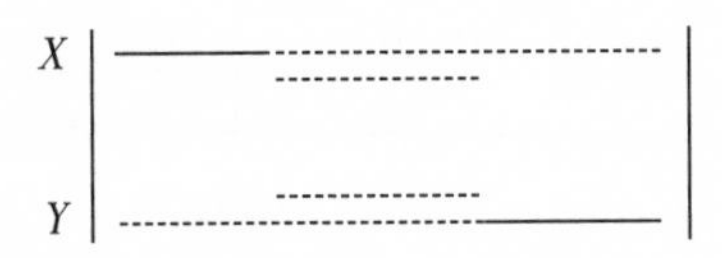

Figure 38. De Morgan's graphical representation of "Some things are neither *X*s nor *Y*s."

of thought to necessary *matter* of thought. . . . A generation will arise in which the leaders of education will know the value of logic, the value of mathematics, the value of logic in mathematics, and the value of mathematics in logic."[12] In this, he was influenced by the genius of George Boole, who had a brilliance and originality that De Morgan was among the first to encourage and recognize.

Leibniz's Dream Comes True: Boolean Logic

Limited to a "National School" education, George Boole's knowledge of mathematics was primarily self-taught. Boole's interest in logic was revived by the public controversy between his close friend and mentor, Augustus De Morgan, and the highly respected Sir William Hamilton. Boole felt that De Morgan was right and Hamilton wrong, and in 1847 the dispute spurred him to publish a short work entitled *Mathematical Analysis of Logic*. The mathematicians of the British school were at the time having an algebra revolution of sorts. Having recognized that algebraic systems need not have anything to do with "numbers," they extended the ideas of algebra to other objects. De Morgan immediately recognized that Boole had achieved a synthesis where no one else had; Boole's analysis exhibited the laws of thought in logic in a form as rigorous and exact as any in pure mathematics.[13] Leibniz and De Morgan had seen the connection between algebra and logic, but Boole was able to create the laws of a symbolic algebraic logic, a mathematical logic. And the laws of Boole's system were surprisingly simple—simple enough to be relayed by the electrical impulses of a computer down the road. Logic had been reduced to an extremely easy type of algebra, which came to be called *Boolean algebra*.

Over the next few years, Boole developed and polished his algebra of logic, which led to his 1854 masterpiece, *An Investigation into the Laws of Thought*. Boole's variables stood for classes—we needn't specify what is in the classes. They could be classes of things, ideas, numbers, geometrical objects, or propositions. Boole introduced the universe class (like De Morgan's *universe of discourse*) of which all things are members, represented by the number *1*—100 percent of everything under discussion—and the empty class of which nothing is a member, represented by the number *0*—0 percent of everything under discussion.[14] If ***X*** and ***Y*** are classes, ***X*** equals ***Y*** (***X*** = ***Y***) means that the classes have the same members.

Furthermore, we can define operations between the classes. The class defined by ***1*** − ***X*** is the class of everything in the universe after ***X***-things are removed—that is, the class of **not-*X*** things (De Morgan's negative term). The *logical conjunction*, ***X*** *and* ***Y***, came to be called the *logical product* and was symbolized as ***X*** × ***Y***, or ***XY***. ***XY*** is the class consisting of things that belong to both class ***X*** and class ***Y*** and is also called the intersection, or overlap. If ***X*** represents the class of yellow things and ***Y*** represents the class of hairy things, then ***XY*** represents the class of yellow, hairy things. One immediate result, no matter what class ***X*** represents, is that the logical product ***1X*** equals ***X***—meaning that the class consisting of things that belong to both the class of everything (universe class) and the class of ***X***-things is equal to the class of X-things. A second result, regardless of the membership of X, is $0X = 0$—meaning that the class of things that belong to both the class of nothing (null class) and the class of X-things is equal to the class of nothing. It is extremely appealing to have the logical product $1X$ equal to X and the logical product $0X$ equal to 0—exactly like our algebra of numbers. But unlike our algebra of numbers, in Boole's algebra, $XX = X$. However, it makes sense that the members of both X and X are simply the members of X. Logical *disjunction* was defined as the *logical sum*, $X + Y$, meaning X *or* Y (also called the *union* of X and Y).[15] This gives way to two other rather jarring rules to the beginning algebra student: $1 + 1 = 1$, since the union of all things and all things only equals all things (not more) and, as it has evolved to modern day, $X + X = X$.

The law of the excluded middle could be formulated as an *equation*: $X(1 - X) = 0$, meaning "The class of things common to both the class of X-things and the class things **not-X** is the class with no items in it." Or, as Aristotle had stated, it is impossible that the same thing both is and is not at the same time. Aristotle's four categorical propositions were formulated as equations as shown in Table 3.

Table 3. Boole's Symbolic Logic

Aristotle's Propositions	Boole's Equations
A: All X is Y.	$X(1 - Y) = 0$
E: No X is Y.	$XY = 0$
I: Some X is Y.	$XY \neq 0$ or $XY = V$* where $V \neq 0$
O: Some X is not Y.	$X(1 - Y) \neq 0$ or $X(1 - Y) = V$ where $V \neq 0$

*Boole preferred to express all traditional categorical propositions as equations rather than as inequalities.

Furthermore, in Boole's system of algebraic logic we will allow the symbols *x*, *y*, and so on, to take on the values of 0 and 1, where $x = 1$ means proposition ***X*** is true and $x = 0$ means proposition ***X*** is false. Nowadays we call this the "truth value" of a proposition, and the truth value of any more complicated statements can be determined by numerical computation following Boole's few rules. This convention is used today by digital computers wherein statements that are true are assigned the logical value 1 and statements that are not true (they are false) are assigned the value 0. With the exception of $1 + 1 = 1$, the rules for disjunction follow the arithmetic rules of addition and the rules for conjunction follow the arithmetic rules of multiplication. Rules for disjunction, conjunction, and the conditional are shown in Table 4. These rules state that a disjunction is false if and only if both constituent propositions are false; a conjunction is true if and only if both constituent propositions are true; and a conditional is false if and only if the antecedent is true and the consequent is false.

An important addition to Boole's laws of symbolic logic are two rules creating what is called a *duality* between disjunction and conjunction. Rediscovered in the nineteenth century, they have become known as "De Morgan's Rules," but the rules were understood far earlier. They were among the works of William

Table 4. Rules for Computing Truth Values

Disjunction	Conjunction	Conditional
$1 + 1 = 1$	$1 \times 1 = 1$	If 1, then $1 = 1$
$1 + 0 = 1$	$1 \times 0 = 0$	If 1, then $0 = 0$
$0 + 1 = 1$	$0 \times 1 = 0$	If 0, then $1 = 1$
$0 + 0 = 0$	$0 \times 0 = 0$	If 0, then $0 = 1$

of Ockham who lived in the early 1300s. Ockham is famous for a maxim known as Ockham's Razor, or Occam's Razor, which states that assumptions introduced to explain a phenomenon should not be multiplied beyond necessity—in other words, all else being equal, the theory with the fewest assumptions is the best.[16] Others have indicated that the Stoics must have realized this connection between negation, disjunction, and conjunction in formulating their inference schema. De Morgan himself states the rules as, "The contrary of an aggregate is the compound of the contraries of the aggregates; the contrary of a compound is the aggregate of the contraries of the components"—that is, the negative of a disjunction is the conjunction of the negations of the disjuncts, and the negation of a conjunction is the disjunction of the negations of the conjuncts.[17] Hence, the propositions "***not***(***X*** or ***Y***)" and "***not***-***X*** and ***not***-***Y***" are equivalent. An example is the equivalence of the statements "It is not the case that you are either a swimmer or a golfer" (we would usually say "You are neither a swimmer nor a golfer") and "You are not a swimmer and you are not a golfer." An example for the second rule, ***not***(***X*** and ***Y***) = ***not***-***X*** or ***not***-***Y***, is the equivalence of "It is not the case that you are both a trumpet player and a trombone player" and "Either you are not a trumpet player or you are not a trombone player."

De Morgan's entry in *logic* in the 1860 *English Cyclopaedia* exemplifies the fusion of algebra and logic that Boole's work initiated. He illustrates how algebraic transformation can be used to show that "to be both ***not***-***A*** and ***not***-***B*** is impossible" is logically equivalent to "every thing is either ***A*** or ***B*** or both." His entry states:

> Let A and B represent two objects of thought. Let 1 represent the universe, all that exists; let 0 represent the impos-

sible, something that does not exist. Let = represent identity. Let A + B represent the class containing both A and B, with all the common part, if any counted twice; let A – B signify what is left of the class A, when B, which it contains, is withdrawn. Let AB represent the common part of the notions A and B. Then 1 – A and 1 – B represent all that is not A and all that is not B; and the non-existence of everything which is both not A and not B is symbolized by

$$(1 - A)(1 - B) = 0.$$

The common rules of algebra transform this into

$$A + B - AB = 1.$$ [18]

This small passage refers the interested student to the work of George Boole. But as the entry is written for the lay reader, it will be interesting for us to try to understand it and appreciate the harmonious way that logic lends itself to algebraic manipulation. ***A*** and ***B*** are Boole's classes. ***A + B*** denotes ***A*** *or* ***B*** *or both* (*both* is the common part to which De Morgan refers), ***A – B*** denotes what is left when all class ***B*** things are removed from class ***A***, and ***AB*** is the class of things that are *both* ***A*** *and* ***B***. **Not-*A*** is represented by ***1 – A***, and **not-*B*** is ***1 – B***. "Both **not-*A*** and **not-*B*** is impossible" is given by the equation ***(1 – A)(1 – B) = 0***. The logical product on the left-hand side is multiplied out precisely as it is in algebra, to transform ***(1 – A)(1 – B) = 0*** into ***1 – A – B + AB = 0***. With a little more algebra, we arrive at ***1 = A + B – AB***, which tells us that everything is either ***A*** or ***B*** or both (and since the common part, ***AB***, is counted twice, we remove one counting of ***AB***).

Boole's mathematical career began late and ended early; he died in 1864, at the age of 49. Over the next half-century, Boole's theory was tweaked and polished. Augustus De Morgan and William Stanley Jevons continued Boole's work in England and Charles Sanders Peirce and his colleagues continued it in the United States.* The German mathematician Gottlog Frege's 1879 treatise *Begriffschrift* and the Italian mathematician Giuseppe Peano's 1894 *Notations de Logique* combined logic with the study of sets and numbers.[19] Peano adopted the symbols for the connectives—*not*, *or*, *and*, *implies*, and *if and only if*—that are almost identical to those used today. From the work of Frege, Peirce, and Peano grew the broader field of logic (including *propositional calculus* and *relations*), which became known as *predicate calculus*.

Boole's ideas were popularized by Bertrand Russell and Alfred North Whitehead in their famous 1910 *Principia Mathematica*. Russell and Whitehead's massive work tried to derive all of mathematics by using the basic laws of predicate calculus. David Hilbert streamlined the work of Russell and Whitehead to develop his concept of "formal systems" in *Uber das Unendliche* in 1926.[20] Some symbols have become standard, and others depend on what convention the author adopts. Table 5 illustrates some of the different symbols used in logic.

*Peirce introduced indices and the summation symbol (Σ) and product symbol (Π) for logical addition and logical multiplication, respectively. He attributes the use of indices into the algebra of logic to Mr. Mitchell (Peirce 1883). Σ meant "some" so that $\Sigma_i x_i$ means that x is true of at least some of the individuals denoted by i; in other words x_1 is true *or* x_2 is true *or* x_3 is true, and so on. Π means "all" and $\Pi_i x_i$ means that x is true of all these individuals; that is, x_1 is true *and* x_2 is true *and* x_3 is true, and so on. For example, if ℓ_{ij} means i loves j and ℓ_{ii} means i loves himself, then $\Pi_i \Sigma_j \ell_{ij} = (\ell_{11} + \ell_{12} + \ell_{13} + \text{etc.}) \cdot (\ell_{21} + \ell_{22} + \ell_{23} + \text{etc.}) \cdot (\text{etc.})$ means everybody loves somebody, but $\Sigma_j \Pi_i \ell_{ij} = \ell_{11} \cdot \ell_{21} \cdot \ell_{31} \cdot \text{etc.} + \ell_{12} \cdot \ell_{22} \cdot \ell_{32} \cdot \text{etc.} + \text{etc.}$ means somebody is loved by everybody (Peirce 1933, vol. 3, pp. 393 and 498–502).

Table 5. Logic Symbols

OPERATION	PEANO-RUSSELL	HILBERT	VARIATIONS
Negation	$\sim p$	$\overline{p}$	$-p, \neg p$
Conjunction	$p \cdot q$	$p \& q$	$pq, p \wedge q$
Disjunction	$p \vee q$	$p \vee q$	pq
Conditional	$p \supset q$	$p \rightarrow q$	$p \prec q$
Biconditional	$p \equiv q$	$p \sim q$	$p \leftrightarrow q$

In much the same way that Aristotle had tried to simplify his own system by reducing all valid syllogisms to *Barbara* or *Celarent*, there have been other attempts to invent new symbols to either reduce or simplify notation. For complex propositions, parentheses were necessary to resolve amphibolies. For example, does "**A** and **B** or **C**" mean "**A** and (**B** or **C**)" or "(**A** and **B**) or **C**"? Parentheses could be minimized if an order of operations was agreed upon but further rules would be required.[21] In 1913, Harvard Professor Henry M. Sheffer invented a single symbol, |, that could do the duty of all the operations: negation, conjunction, disjunction, conditional, and biconditional. Called the Sheffer stroke operation, every statement of logic can be written in terms of this symbol alone.[22] According to Sheffer, $p \mid q$ meant "neither p nor q," represented as $\sim (p \vee q)$ in the Peano-Russell notation. However, when the number of symbols is reduced in an effort to simplify (Ockham's Razor), things get surprisingly complicated. The simple disjunction "p or q," represented as $p \vee q$ in the notation evolved from Boole, looks like $p \mid q \mid p \mid q$ when written in terms of Sheffer strokes. The biconditional "if p then q and if q then p," also expressed as "p if and only if q" and commonly symbolized by $p \leftrightarrow q$, would be expressed as $p \mid p \mid q \mid q \mid q \mid p$ using Sheffer strokes but is rather hard to express in language.

The Sheffer stroke is a building block of digital electronics and it is common today to express the stroke as the "not both" operation, rather than the "neither/nor." Sheffer himself indicated that this was an alternate interpretation due to the dual nature of disjunction and conjunction. In circuit design, the Sheffer stroke operation is called a *NAND*—standing for *Not AND*, and the symbol for the *NOR* operation is called the Peirce arrow. Today, the NAND or NOR operations (called gates) are often used to model the circuitry of electronic devices.

Boolean algebra was used as early as 1936 for the analysis of switch and relay circuits in electrical communication engineering. Today, the computer itself uses the rules of logic in instructing its electrical circuits and networks how to operate. "On" is akin to the affirmation of a true proposition (1) and "off" to the denial of the same proposition (0). Two switches in series behave like the conjunction of two propositions (both must be on), while switches in parallel behave like the disjunction of propositions (one or the other must be on). In addition to electrical energy, Boolean algebra can be applied to any system in which energy is transmitted through conduits that can be switched on or off—light beams, sound waves, fluid control systems, even odor.[23]

Today all computer science students study Boolean logic. In fact, many computerized search systems employ Boolean logic. Arriving as it did long before the first modern computer, the publication of Boole's work prompted several notable attempts to create machines that could make use of logic. Since the laws of logic can be followed without regard to the subject matter being reasoned about, why couldn't a machine perform the arithmetic operations of Boole's algebraic logic? Let us now take a careful look at logic machines.

10

Logic Machines and Truth Tables

The secret of all reasoning machines is after all very simple. It is that whatever relation among the objects reasoned about is destined to be the hinge of a ratiocination, that same general relation must be capable of being introduced between certain parts of the machine.

Charles Sanders Peirce

Reasoning Machines

As logic enjoyed a rebirth through the work of George Boole, there were several notable attempts to mechanize the tedious work of analyzing long syllogisms. One of Professor Augustus De Morgan's many devoted students, William Stanley Jevons, the British logician, philosopher, and economist, caught the attention of the logicians of the time when in 1869 he produced a rather famous Logic Machine.[1]

Jevons was actually preceded in this ambitious endeavor by the British statesman and inventor Earl Stanhope, some 50 years earlier. Stanhope's device—the Stanhope Demonstrator—employed colored sliding panels that one maneuvered into slots according to the premises of the syllogism. The rather simple device could handle only three terms but allowed quantification of the predicate and even numerically definite syllogisms.[2] Jevons could not possibly have known of Stanhope's work as the

Earl was known to be extremely secretive, obsessed with the notion that "some bastard imitation" might precede his intended publication. Sadly, Stanhope died before he could publish news of the Demonstrator. Through accounts garnered from letters, the Stanhope Demonstrator was brought to light later, in 1879, 63 years after his death.[3]

Jevons's Logic Machine, or "analytical engine" as he called it, was the first working model able to solve a complicated syllogism faster than a proficient human being could. Like De Morgan, Jevons was one of the few British logicians to recognize the pioneering aspects of Boole's accomplishments in algebraic logic. Jevons considered his machine to be a purely mechanistic embodiment of Boole's *Laws of Thought*; moreover, his well-known mechanical device established a prototype for those that were to follow.[4] The logic machine was about a meter in height and was sometimes referred to as the "logical piano," resembling as it did, an upright piano. Modern historians have described it as resembling a cash register and indeed it does.[5] The pianolike "keyboard" had 21 "keys" that provided a means for entering the premises as equations. The logic machine was based on Jevons's method of combining terms and could handle syllogisms involving four terms and their negatives, as well as all the logical combinations among the terms. Rather than outputting a conclusion once the premises were input, the mechanism displayed a list of all possible conclusions to be drawn from the premises after having eliminated all contradictory propositions. The user had to inspect the possible conclusions and eliminate those that were not applicable; those remaining (if any remained) were the appropriate conclusions. In 1870, Jevons demonstrated the logical piano at a meeting of the Royal Society of London where the machine brought attention to the value of Boole's symbolic system and to the possibility of "mechanistic thinking."

Jevons called his system of combining terms to represent premises *combinational logic*, and he dubbed an exhaustive index of combinations of a logical alphabet a *Logical Abecedarium*.[6] Jevons went so far as to calculate the amount of space necessary to house a complete index for five terms. With each page displaying 64 entries and each volume containing 1,024 pages, the index would require 65,536 such volumes. His calculations are reminiscent of the enumerations made by Chrysippus and Hipparchus, thousands of years earlier. At the time, these amounts must have seemed overwhelming, but modern computer storage devices can handle them easily.

As early as 1863, Jevons had invented a Logical Slate, which consisted of a complete abecedarium that was permanently engraved upon a common writing slate. Jevons also suggested several labor-saving devices such as the creation of a rubber stamp of the logic alphabet, thus eliminating the tedium of having to write down all of the combinations every time. For classroom instruction, Jevons favored a device he created called a Logical Abacus in which the combinations of the abecedarium were written on movable strips of wood, and a syllogism was analyzed by the user manually performing procedures similar to the ones his machine performed.[7]

Jevons's method was actually worked out before John Venn's method of diagrams, but Venn considered his diagrammatic analysis to be much easier than organizing premises into the form appropriate for Jevons's machine. In 1880, Venn proposed a jigsaw puzzle version of his five-term diagram with each of the 32 compartments represented by a puzzle piece that would be removed as possibilities were eliminated. In addition, he designed what he termed a "logical-diagram machine," which was a three-dimensional version of the jigsaw puzzle with four overlapping elliptical cylinders.[8] Although they argued over the rival notions of whether diagrams or machines were more

straightforward, Jevons and Venn clearly intended their contributions to elucidate the genius of Boole's system. Their contributions were merely the icing on the cake; the cake itself was created by George Boole in the two books he wrote in 1847 and 1854.

In 1881, the American Allan Marquand constructed a logical-diagram machine, an improved version of the Jevons machine. The Marquand Logic Machine was smaller, the number of keys reduced, and its system of rods, springs, and levers proved far more sophisticated than previous mechanisms.[9] The control panel of the instrument resembled Marquand's logic diagram of overlapping rectangles.

Although it performed the same operations as the Jevons-type machine, Marquand felt that his device could be easily adapted to solve much larger problems.[10] Both Jevons and Marquand had used De Morgan's negative terms (represented by lowercase letters) as inputs to their machines. One interesting aspect of Marquand's machine is that entire premises of the syllogism had to be input *in the negative*. His machine then eliminated any conclusions that agreed with the negative of the premises, as those would contradict the premises. In an 1885 article for the *Proceedings of the American Academy of Sciences*, Marquand described his machine and included pictures of it. He used Boole's notion of disjunction as logical addition and conjunction as logical multiplication and Peirce's symbolic notation for the conditional.[11] $A \prec B$ means "If ***A***, then ***B***" (or "Every ***A*** is ***B***" or "Class ***A*** is included in class ***B***"). To input this premise, Marquand explained that we input its negative, $Ab \prec 0$—meaning "***A***s that are ***b*** (not-***B***) do not exist."

Charles Sanders Peirce described the logic machines built by Jevons and Marquand as mills into which the premises were fed and conclusions turned out by the revolution of a crank.[12] Peirce commented:

> Precisely how much of the business of thinking a machine could possibly be made to perform, and what part of it must be left for the living mind, is a question not without conceivable practical importance; the study of it can at any rate not fail to throw needed light on the nature of the reasoning.[13]

The machines of Jevons and Marquand, while utilized to solve problems of a fairly elementary nature, afforded the world evidence of the possibilities of a reasoning machine that employed the rules of logic.

In the 1950s, a wiring diagram for a four-term electrical logic machine was found among Marquand's manuscripts. Believed to be prepared in 1885, it is probably the first circuit diagram of an electric logical machine. Marquand had been a student of philosopher and Harvard professor Charles Sanders Peirce, and in the early 1970s an extremely interesting letter came to light that Peirce had sent to Marquand in 1886. In the letter, Peirce suggested a method in which Marquand's machine might be improved by using electricity. Peirce even produced a sketch detailing how circuits for logical conjunction and disjunction would perform in series and parallel. Prolific author of books in recreational mathematics, Martin Gardner says that this is "the first known effort to apply Boolean algebra to the design of switching circuits!" in the same way that modern computer designers do today.

Gardner has written a delightful book, *Logic Machines and Diagrams*, which chronicles the progress of syllogism machines. He points out that ironically none of the syllogism machines of the time actually used logic to solve syllogisms. Even when electrical connections were introduced into the instruments, the electrical connections were not controlled through logical translations as they are today. The syllogism machines merely used electricity to reveal or cover up a preset arrangement of valid or contradictory statements rather like the windows and levers of the original mechanical devices. The link between Boolean algebra and switching circuits had yet to be made, except perhaps by Marquand and Peirce.

By the 1930s the link was indeed made, and since then hundreds of papers applying logic to switching circuits have been written. One other logic machine built by two Harvard undergraduates in 1947 is rather interesting. William Burkhart and Theodore A. Kalin, who were taking a course in symbolic logic from renowned logician Willard V. Quine, constructed the first electrical machine designed exclusively for propositional logic for the sole purpose of doing their homework problems. The Kalin-Burkhart "logical truth calculator" could handle up to 12 terms by isolating lines in a truth table.

Truth Tables

The "truth table method" was introduced in 1920 in the Ph.D. dissertation of Emil Leon Post, a young Polish, Jewish emigrant student at the College of the City of New York. Used extensively to this day in the study of logic, a truth table is a table of all possible combinations of true/false (or 1s and 0s) for the propositions involved in an argument, using the rules set down by Boole.[14] Earlier, we constructed a truth table for disjunctive, conjunctive, and conditional statements involving two propositions. Using a truth table, we can answer a problem adapted from Allan Marquand:

> Suppose, regarding three girls, Anna, Bertha, and Cora, we observe the following rule:
>
> Whenever either Anna or Bertha (or both) remained at home, Cora was at home; and

When Bertha was out, Anna was out; and
Whenever Cora was at home, Anna was at home.
What can we determine about the habits of the girls individually or as a group?[15]

Letting ***A***, ***B***, and ***C*** stand for "Anna is at home," "Bertha is at home," and "Cora is at home," respectively, we must judge the truth of propositions ***A***, ***B***, and ***C*** given the truth of the rule "If ***A*** or ***B*** then ***C***, *and* if **not-*B*** then **not-*A***, *and* if ***C*** then ***A***." A truth table consists of an exhaustive list of possible truth values. In our example, there are three simple propositions, ***A***, ***B***, and ***C***, each with two possible truth values (true or false). A truth table to analyze the rule involving all three propositions would require 2 × 2 × 2, or 8, lines in the table to reflect all true/false combinations of ***A***, ***B***, and ***C***. The rule is a conjunction of three compound propositions, which must each be true for the whole rule to be true, since they are connected by "and." Working with the first of the three conjoined propositions, "If ***A*** or ***B*** then ***C***," we get:

A	*B*	*C*	*A* OR *B*	IF *A* OR *B* THEN *C*
T	T	T	T	T
T	T	**F**	**T**	**F**
T	F	T	T	T
T	F	**F**	**T**	**F**
F	T	T	T	T
F	T	**F**	**T**	**F**
F	F	T	F	T
F	F	F	F	T

"***A*** or ***B***" is true any time either ***A*** is true or ***B*** is true and false only when both are false. "If ***A*** or ***B*** then ***C***" is always true except when the consequent (***C***) is false and the antecedent (***A*** or ***B***) is true. Those have been highlighted in the table. Now since

"If ***A*** or ***B*** then ***C***" must be true for the entire rule to be true, we can eliminate any line in which "If ***A*** or ***B*** then ***C***" is false (has an F). We will therefore eliminate those three lines from the table.

Let's add the second part of the rule, "if ***not-B*** then ***not-A***." Before we do, however, let's add a column for ***not-B*** and ***not-A***, so the analysis will be easier. ***Not-B*** receives a value "F" whenever ***B*** has a "T" and a value "T" whenever ***B*** has an "F." The same holds for ***not-A*** with respect to ***A***. "If ***not-B*** then ***not-A***" is false whenever ***not-B*** is true and ***not-A*** is false.

A	***B***	***C***	If ***A*** or ***B*** then ***C***	***Not-B***	***Not-A***	If ***not-B*** then ***not-A***
T	T	T	T	F	F	T
T	F	T	T	**T**	**F**	**F**
F	T	T	T	F	T	T
F	F	T	T	T	T	T
F	F	F	T	T	T	T

Let's eliminate the line where "if ***not-B*** then ***not-A***" is false (highlighted) and add the last portion of the three-part rule, "if ***C*** then ***A***."

A	***B***	***C***	If ***A*** or ***B*** then **C**	If ***not-B*** then ***not-A***	If ***C*** then ***A***
T	T	T	T	T	T
F	T	**T**	T	T	**F**
F	F	**T**	T	T	**F**
F	F	F	T	T	T

If we eliminate the lines where "if ***C*** then ***A***" is false, we are left with only two lines in the truth table where all three parts of the rule are true. What does the truth of the rule indicate about the truth of ***A***, ***B***, and ***C*** individually or as a group? Looking back to the first three columns and the truth values of ***A***, ***B***, and ***C***, we are able to see that all must be true or all must be false. We can conclude that if the rule is true, then either all three girls are at home at the same time or all three are away at the same time. The fact that Marquand's machine required the premises to be input in the negative is interesting because it is so similar to looking at a truth table and eliminating the false scenarios.

True, False, and Maybe

In our analysis of Marquand's problem about Anna, Bertha, and Cora, each statement, simple or compound, was either true or false—the traditional system of logic. The law of the excluded middle guarantees us that a proposition is always either one or the other, and the law of noncontradiction guarantees us that a proposition is never both. In the language of the logician, the logic is *bivalued*, meaning that each proposition has one of two truth values. Logics other than the classical logic have been developed that do not restrict the number of truth values to two; in fact they may allow for a much larger set of truth values.

In 1917, Jan Łukasiewicz, co-founder of the Warsaw School of Logic and on whose work the great mathematical logician Alfred Tarski based his own, advocated a system with a third truth value, "possible." Aristotle himself had acknowledged that inferences are often drawn from premises such as "It may be that all are . . ." or "Some possibly are. . . ." In a theory of what is called *modal logic*, Aristotle attempted to develop the same systematic treatment of syllogisms involving statements of necessity, state-

⊃	Then 1	Then ½	Then 0
If 1	1	½	0
If ½	1	1	½
If 0	1	1	1

Negation	~
Not 1	0
Not ½	½
Not 0	1

Figure 39. Truth values for the conditional and negation under Łukasiewicz's three-valued logic.

ments of actuality, and statements of possibility, but he was never able to devise an organized and satisfactory system for modal propositions in the way that he had for categorical propositions. Prompted by the idea of understanding the modal notion of *possibility* in a three-valued way, Łukasiewicz suggested a three-valued logic with propositions categorized as "true," "false," or "possible."

Łukasiewicz proposed that if the certainty of a premise could not (yet) be established, that is, assigned a truth value of 1 (it is necessary) or assigned a truth value of 0 (it is impossible), we may indicate "the possible" by a truth value of ½. See his truth values for the conditional and for negation in Figure 39.[16]

The truth values for the conditional and the negation of true (1) and false (0) statements are the ones used in standard logic. Clearly the negation of "possible" is not "impossible" but "possibly not." Take, for example, the proposition "You will win the lottery"—a proposition whose certainty will only be known in the future. Right at this moment, it is *possibly true*. Its negation is "You will not win the lottery" and it is also *possibly true*. Under a system of three-valued logic, both propositions will receive a truth "value" of ½. It may seem strange that both propositions receive a value of ½, since the chances of your not winning the lottery are greater than the chances of your winning, but we will return to that issue later. In classical logic no formula can be equivalent to its own negation; but in a three-valued logic if ***A*** has a value of ½, so does ***not-A***. Therefore, ***A*** and ***not-A*** can be considered equivalent. The introduction of three truth values produces some surprisingly counterintuitive results. Our most basic laws of logic, the law of the excluded middle and the law of noncontradiction, are violated in a three-valued logic system.

Some modern linguists and philosophers have preferred three truth values for different reasons.[17] A statement can be true, false, or neither true nor false; a statement is assigned the third truth value in cases where an *existential presupposition is violated*. How does one assign a truth value to a statement like "The present king of France is bald"? If we presuppose that the proposition actually speaks of something, we are mistaken. Our supposition is violated; there is no king of France. France no longer has a monarchy. Perhaps we should declare the proposition to be false. If there is no king of France then he can't possibly be bald. However, if the proposition is false then its negation must be true. We would be forced to accept the truth of "The present king of France is not bald," but once again we find ourselves in a quandary.[18] Some have preferred to assign a third truth value to propositions of this sort (those that are neither true nor false). Statements that violate our existential presupposition do not fit neatly into Aristotle's framework, and it is not clear how to treat them in logic. Forming the negations and understanding the truth rules of conditionals formed from such statements prove to be even more problematic.

According to the *Stanford Encyclopedia of Philosophy*, around 1910, Charles Sanders Peirce had developed, but not published, what amounts to a three-valued logic. Peirce used three symbols ***V***, ***L***, and ***F***; ***V*** was associated with true (1), ***F*** was associated with false (0), and ***L*** was associated with an intermediate or unknown value (½).[19] Peirce defined the rules for operating with three truth values: negation, disjunction, and conjunction, as well as inventing some operators of his own. Other philosophers, mathematicians, linguists, and psychologists have made forays into the usefulness of a three-valued logic. The third truth value has been given various interpretations, such as "undefined," "senseless," "undetermined," or "paradoxical."

Four-valued systems have been developed that have applications in computer science[20] and in 1921, Emil Leon Post published "A General Theory of Elementary Propositions" in *The American Journal of Mathematics*, wherein he proposed many-valued systems of logic. Generalizing his Ph.D. thesis, Post produced a framework for a system of logic based on an arbitrary but finite number of truth values, rather than the two truth values of true and false.

Attempts to merge modal logic and many-valued logic have applications to problems dealing with artificial intelligence, a field in which scientists are trying to model human thinking. Many-valued logics have applications in linguistics, applications to philosophy in resolving certain paradoxes, applications in mathematics, and applications in hardware design. In the same way that classical logic is used as a technical tool for the analysis of electrical circuits built from switches with two *states*—"on" and "off"—a many-valued logic may be used to analyze circuits built from switches with many states.

Today machines can perform many amazing tasks; they can apply the rules of a two-valued logic or a many-valued logic unfailingly and mechanically. However, Charles Sanders Peirce's words written over a hundred years ago still ring true:

> Every reasoning machine, that is to say, every machine, has two inherent impotencies. In the first place, it is destitute of all originality, of all initiative. It cannot find its own

> problems; it cannot feed itself. . . . This, however, is not a defect in a machine; we do not want it to do its own business, but ours. . . . In the second place, the capacity of a machine has absolute limitations; it has been contrived to do a certain thing, and it can do nothing else. . . . But the mind working with a pencil and plenty of paper has no such limitations.[21]

Mr. Peirce is right. Humans are superior to machines in their creative abilities and initiative. They are superior in their inherent wherewithal to manipulate concepts that are not quite black and white. And most concepts are not black and white; most concepts (as we'll see next) are fuzzy.

11

Fuzzy Logic, Fallacies, and Paradoxes

Shaggy logic underlies both ordinary language and most human thinking.

Martin Gardner

Shaggy Logic

We often hear those in authority criticized for their "fuzzy math" when their data are dubious or their figures are suspect. In this context, the "fuzzy math" remark is disparaging the accuracy of the data or the soundness of the methods used to arrive at the figures. But make no mistake, fuzzy logic is not "fuzzy" in that sense. Fuzzy logic is a logic of fuzzy concepts, it is not a logic that is itself fuzzy.[1] Fuzzification takes into account the imprecision of data, the vagueness of language, and the uncertainty inherent in systems. Whereas two-valued Boolean logic is sufficient for worlds with two states such as true or false, off or on, and black or white, fuzzy logic allows us to deal with shades of gray.

An extension of many-valued logic, fuzzy logic is an attempt to assign truth values to concepts to handle partial truth and uncertainty. A completely true proposition would be assigned a truth degree 1 and a completely false proposition would receive a truth degree 0. If neither 100 percent true nor 100 percent false, a proposition would receive a value between 0 and 1 depending on its degree of truth. The closer a proposition is to being completely true, the closer its truth degree is to 1, and the closer it is to being completely false, the closer its truth degree is to 0. Some propositions are simply truer than others. "Albert Einstein is (or was) smart" and "Bart Simpson is smart" may both be true statements. One may be truer than the other.

The concepts of truth in formal logic, like the on/off switches of an electrical circuit, are said to be *crisp*. But surely most concepts are fuzzy, not crisp, and we manipulate them handily in the course of ordinary thought. If a concept is extremely fuzzy, it is called *vague*. Martin Gardner cites the examples: "Bob will be back in a few minutes" is fuzzy, but "Bob will be back sometime" is vague.[2]

In the way that standard logic is associated with sets and class membership, fuzzy logic is akin to the concept of fuzzy set theory. The term "fuzzy logic" materialized during the development of the theory of fuzzy sets, which was pioneered in the United States by Lotfi Zadeh in 1965. Fuzzy set theory was introduced as a mathematical approach to account for the imprecision of everyday life. The members of fuzzy sets are characterized by a *degree of membership* to the set, and the propositions of fuzzy logic are characterized by their *degree of truth*. Rather than every proposition being absolutely true or absolutely false, a proposition may be judged absolutely true, absolutely false, or some *intermediate* truth value, representing its degree of truth.

In fuzzy logic a truth value is assigned to propositions like "Mary is tall" or "Mary is smart." But, how tall does one have to be to be considered "tall"? And, how smart does one need to be to be considered "smart"? If our universe of discourse is the set of "tall people," Mary can be assigned a number between 0 and

1, depending on her degree of membership in that set. A 1 indicates complete membership to the set, and a 0 indicates nonmembership. If Mary is seven feet tall, she will be assigned a 1—most definitely she is a member of the set of "tall people." The proposition "Mary is tall" is completely true and we can use the rules of ordinary Boolean logic. If she is four feet tall, Mary will be assigned a 0—she's not a member of the set. The proposition "Mary is tall" is false and once again we have reverted to Boolean logic. If Mary is 5' 5" tall (1.651 meters), the proposition "Mary is tall" has a degree of truth and Mary has a degree of membership in the set of tall people. Mary will be assigned a number between 0 and 1 that quantifies the degree to which she is a member of the set of tall people. Compound propositions in fuzzy logic can then be valuated by consulting the truth degrees of the constituent propositions and applying the rules of symbolic logic to the connectives in the same way that we do in standard logic.

There is no single interpretation of truth degree; how it is to be interpreted depends on the field of application. In capturing the vagueness and imprecision of natural language concepts like tallness and smartness, truth degree can measure the degree of similarity. As early as 1704, Gottfried Leibniz had suggested using probabilities to measure degrees of truth and commented that Aristotle himself had held the opinion that inferences can be made from the probable.

> Opinion, based on the probable, also deserves perhaps the name knowledge; otherwise nearly all historical knowledge and many other kinds will fall. But without quarreling over names, I hold that *the investigation of degrees of probability* would be very important, that we are still lacking in it, and that this lack is a great defect of our Logic.

> For when we cannot decide a question absolutely, we might still determine the degree of likelihood from the data, and can consequently judge reasonably which side is the most likely.
>
> I have more than once said that we should have a *new kind of Logic* which would treat of degrees of probability, since Aristotle in his *Topics* has done nothing less than that.[3]

Probabilities can be utilized to determine membership degree but degrees of membership and probabilities are not equivalent. Fuzzy expert James Bezdek gives the example of two bottles of liquid offered to you after a week in the blazing hot desert with nothing to drink. Bottle **A** has been assigned a degree of membership of 0.90 to the class of potable liquids, and bottle **B** has been assigned a probability of 0.90 of being potable. Which do you drink? The 0.90 membership degree is interpreted as the degree to which bottle **A** is *similar* to the most perfect potable liquid—say, pure mountain spring water. Bottle **B**, on the other hand, has been found to be potable 90 percent of the time (and 10 percent of the time possibly deadly). Perhaps bottle **B** was selected at random from among 10 bottles, 9 of which were pure water and 1 of which was poison. With a 0.90 membership indicating how similar bottle **A** is to being 100 percent pure, you can be guaranteed that the impurities are few. On the other hand, with bottle **B** you have a 90 percent chance of getting pure water and a 10 percent chance that bottle **B** is poison. Obviously, you pick bottle **A**. Probabilities are about the likelihoods of outcomes while fuzzy logic is about degrees of truth depending on the similarities of objects to imprecisely defined concepts.[4]

Fuzzy expert systems are the most common users of fuzzy logic; they employ fuzzy logic in system diagnosis, image processing, pattern recognition, financial systems, and data analysis. In expert systems, both vague notions and commonsense reasoning are modeled via fuzzy sets and fuzzy logic. Another application of fuzzy logic occurs in controllers—devices that make adjustments to a system be it mechanical, chemical, electrical, or a combination thereof. A controller can be any type of apparatus such as the thermostat in your home or the cruise control in your car. Fuzzy controllers are used widely in Japanese consumer products such as refrigerators, washing machines, TV camcorders, air conditioners, palm-top computers, automobiles, cameras, robots, and high-speed trains.

There are, however, some difficulties with the laws of fuzzy logic as they are applied to everyday reasoning. Not all natural language sentences are comparable as to their degrees of truth. Even the same sentence in different contexts might have different truth degrees. "The dress is red" might receive a truth degree of 70 percent on the redness scale if the dress is seen in a white room but a truth degree of 20 percent if seen in a rack among many other red dresses. If propositions are no longer simply true or false but shades of gray, how shall the degrees of truth be determined?

There are other uncomfortable repercussions in fuzzy logic. One of our most basic laws in logic and language is violated by fuzzy logic. The law of the noncontradiction no longer holds; a person can be both tall (to a degree) and not tall (to a degree). How are we to evaluate the counterintuitive from the fallacious?

Fallacies

"There exist both reasoning and refutation which appear to be genuine but are not really so."[5] Thus begins *On Sophistical Refuta-*

tions. Once again, the Sophists have provoked Aristotle, and this time they have provoked him into instructing us in how to detect incorrect reasoning. Here, he enumerates the fallacies associated with invalid arguments so that we will know how to refute them and not be taken in by them.

Aristotle classified fallacies into two main categories, those that depended on the particular language employed and those that did not. One Aristotle translator commented, "Some of these fallacies would hardly deceive the most simple minds; others, which Aristotle seems to have been the first person to expose and define, are capable not only of deceiving the innocent but also of escaping the notice of arguers who are employing them."[6]

Fallacies that depend on language are fallacies only in the sense that they produce the false appearance of an argument. They often employ *equivocation*, using words with double meanings or words that take on different meanings depending on the phraseology.[7] They hinge on the ambiguity of language; they are sophistical booby traps. The series syllogism, "A ham sandwich is better than nothing; nothing is better than eternal happiness; therefore a ham sandwich is better than eternal happiness," is an example of a fallacy of language. The word "nothing" is used in two different ways—meaning "zero" in the first premise and "It is not the case that something is better than eternal happiness" in the second.

A search on the Internet for "fallacy" will provide a host of sites replete with lengthy lists of both types of fallacies, accompanied by examples of their use. There is the fallacy of assuming the point that one seeks to prove (*petitio principii*), sometimes called *begging the question*. One can detect its use when, in demonstrating a certain point that is not self-evident, the individual guilty of this offense assumes the point itself (although it is usually disguised). There is the fallacy of argument *ad hominem*, an argument that introduces irrelevant personal circumstances about the opponent or flat out attacks the opponent rather than attacking the argument itself.

The logical reasoning questions on major exams evaluate a test taker's ability to understand, analyze, criticize, and complete arguments. The questions may require identifying assumptions in an argument, drawing a valid conclusion from the argument, and detecting fallacies in the argument. A reasoning question similar to these with an ad hominem argument from a letter to the editor is shown in Figure 40.

Can you identify the ad hominem argument? The correct

Dear Editor: I feel obliged to comment on the unfair review you published last week written by Robert Duxbury. Your readers should know that Mr. Duxbury recently published his own book that covered the same topic as my book, which you asked him to review. It is regrettable that Mr. Duxbury should feel the need to belittle a competing work in the hope of elevating his own book.

The author of the letter above makes her point by employing which method of argument?

A. Attacking the motives of the author of the unfavorable review.

B. Attacking the book on the same topic written by the author of the review.

C. Contrasting her own book with that written by the author of the review.

D. Questioning the judgment of the author of the unfavorable review.

E. Stating that her book should not have been reviewed by the author of a competing work.

Figure 40. Reasoning question involving an ad hominem argument.

answer is A. The author does not attack the reviewer's book (B), nor does she contrast her own (C). She *may* in fact question Duxbury's judgment and believe the author of a competing work should not be the reviewer, but she makes her point by attacking his motive—a desire to promote his own work.

The fallacies of *appeal to force* or *appeal to the multitude* are similar in that their premises rely on accepting a position that is powerful or popular. *Appeal to authority* similarly establishes the strength of an assertion on an authority but one who is not qualified to lend weight to the current argument. *Argument from ignorance* (*ignoratio elenchi*) is ignorance of that which ought to be proved against an adversary. The culprit sometimes shifts the burden of proof. Something is true because it can't be proven false, or something is false because it can't be proven true. An argument from ignorance can be a *red herring argument*, or one that distracts the audience from the issue in question through the introduction of some irrelevancy.

Genuine fallacies, on the other hand, are fallacies that are independent of language and that violate the laws of reasoning.[8] Fallacies in categorical or propositional syllogistic reasoning fall into the category of *non sequitur* arguments, arguments whose conclusions do not follow from their stated premises. You will recall that there are 256 different ways that the **A**, **E**, **I**, and **O** propositions can be combined into three-line categorical syllogisms (of four different figures), and only a few of these forms are valid. The syllogisms that violated certain rules of form have been given names; the fallacies of the undistributed middle, illicit major, illicit minor, positive conclusion with negative major, positive conclusion with negative minor, and positive premises with negative conclusion were all fallacies that dealt with the *form* of the invalid syllogism.

Aristotle identified the fallacy of *conversion*, and it must have been an error as common in 350 B.C. as it is today. Aristotle noted, "The refutation connected with the consequent is due to the idea that consequence is convertible. For whenever, if ***A*** is, ***B*** necessarily is, men also fancy that, if ***B*** is, ***A*** necessarily is."[9] John Stuart Mill, writing in 1874, commented that the error of converting a universal, "All ***A*** are ***B***, therefore all ***B*** are ***A***," is extremely frequent as is the error of converting a conditional proposition.[10]

Charles Sanders Peirce indicated that "faulty thinking" in logic may belong to the domain of the psychologists, and indeed along the way we have identified a number of theories that attempt to explain why individuals persist in converting a conditional: the use of abstract material in experiments, the difficulty of the tasks involved, the difficulty of negation, implicit assumptions, and the interpretation of the conditional as the biconditional.

In her 1962 study, Mary Henle suggested that errors in deduction are not necessarily an indication that we are not logical.[11] Henle's research involved the evaluation of the logical adequacy of deductions in the context of everyday problems, and she was able to identify several processes that interfere with our ability to reason: the failure to accept the logical task, the restatement of a premise or conclusion so that the intended meaning is changed, the omission of a premise, and the introduction of outside knowledge as a new premise. Failure to grasp or accept the logical task means failure to distinguish between a conclusion that is logically valid and one that is factually correct. Errors arise from a failure to grasp the concept of logical validity or the inability to distinguish logical validity from factual status. When subjects arrive at fallacious conclusions or fail to detect a fallacy, they do so because they have undertaken a task different from the one intended and have undertaken it with different premises. Henle concluded that most of us do not approach a reasoning task like a logician and, subsequently, the errors we make do not constitute evidence of faulty reasoning but are rather a function of our understanding of the task. Many fallacies are produced not by faulty reasoning but by specific changes the reasoner brings to the material being reasoned about.

In studies since Henle's, there have been attempts to make certain that subjects attend to the logical task by providing incentive for high performance, such as monetary incentives for arriving at correct answers. Money notwithstanding, these incentives have failed to improve the subjects' performance on such tasks.[12] If they fail to accept the logical task even when they are getting paid, it is more likely that subjects do not grasp the logical task. I try two questions on my university colleague: "All chairpersons must report to the meeting. You are not a chairperson. Do you have to attend?" No, she says. "All math professors work hard. You are not a math professor. Do you work hard?" Yes, she says. She is applying practical logic and supplying premises to the questions. She is using her common sense: She probably (only probably) doesn't have to attend the meeting and she does work hard.

Henle's study also suggested that a strong attitude or emotional involvement with the particular material interferes with the subject's ability to distinguish between drawing a conclusion that is logically valid and one that is believed to be factually correct. "The more personally relevant the material employed, the more difficult it will be to accept the logical task."[13] Sometimes working with familiar material interferes with our ability to stay on the logical track.

There have been mixed results on the question as to why content affects the rules of reasoning. The data are inconsistent, and the issue has not been resolved. But we can agree with researchers when they conclude that different principles govern our reasoning when we deal with familiar as opposed to unfamiliar material.[14]

The difficulty involved in making inferences is not only a function of the content of the problem but also the *form* of the problem. We have seen that certain forms of inference, like modus ponens, are very easy to grasp, even for young children. Other kinds of inference can prove to be quite difficult, even for adults.[15] But investigators tell us that errors in logic account for only part of the errors in deductive reasoning. The difficulty of a reasoning problem is a function of the number and difficulty of the reasoning steps required for us to solve it. The restrictions of our short-term memory or available computing space limit the complexity of deductions that can be done "in the head." As a result, it is often difficult to keep track of information, organize it, and retrieve information from memory.[16]

In the process of evaluating the validity of a syllogism, subjects tend to search for verification. Is there an example in which the premises can be combined to verify a conclusion? Of course, the crucial test is falsification. The reasoner must search his memory for an instance where combining the premises can render the conclusion false. If such an instance exists, then the inference is invalid. If there is no such instance, then the inference is valid. Being able to discover a counterexample or alternative that blocks an unwarranted (yet invited) inference is a valuable skill to have in one's reasoning repertoire.[17] Several researchers have used the following example to illustrate:[18]

All football players are strong.	*versus*	All oak trees have acorns.
This man is strong.		This tree has acorns.
Is he a football player?		Is it an oak?

Because the first syllogism readily invites a counterexample (most folks can think of someone strong who is not a football player), the correct response of "cannot be determined" is elicited. However, in the second syllogism, an example of a tree with acorns that is not an oak is not easily recalled in our memory to provide a counterexample. Logically speaking, the correct response should also be "cannot be determined," but most people will insist that it must be an oak. If you are beginning to think that some of these tasks are deliberately tricky, I agree. My dictionary gives the one and only definition of an acorn as the fruit or seed of an oak tree. No wonder most people are hard-pressed to come up with a counterexample.

A typical error in reasoning is accepting any true statement as logically following from an argument.[19] In formal logic, we carefully distinguish between "truth" and "validity," but in everyday reasoning we are not aware of this distinction. We confuse truth with validity or justifiability. We fall into Henle's categories of "refusing to grasp the logical task" and "the introduction of outside knowledge."

When they aren't sure of the truth of the conclusion, individuals have a tendency to accept any conclusion proffered or invited. Individuals want to arrive at a conclusion. Studies have shown us that time after time subjects are averse to adopting the "no conclusion is possible" or "cannot be determined" position. Specifically because subjects have a strong bias against "can't tell" responses, they often need to be trained and encouraged to use "can't tell" where appropriate. Only then can experiments produce useful results.[20]

Propositional syllogisms that affirm the consequent or deny the antecedent have a correct conclusion of "can't tell." These "can't tell" responses are difficult to elicit from subjects and consequently we find that the syllogisms produce the extremely common fallacies of affirming the consequent and denying the antecedent.

Figure 41 gives another question similar to the ones found in logical reasoning sections of major exams. This time the test taker is required to identify the fallacy.

Here we see Lou making the fallacy of denying the antecedent and Evelyn, who detects a fallacious argument, makes an equally fallacious conclusion. We "can't tell" whether Evelyn's conclusion is true or false; we can simply determine that it is not

---

Lou observes that if flight 409 is canceled, then the manager could not possibly arrive in time for the meeting. But the flight was not canceled. Therefore, Lou concludes, the manager will certainly be on time. Evelyn replies that even if Lou's premises are true, his argument is fallacious. And therefore, she adds, the manager will not arrive on time after all.

Which of the following is the strongest thing that we can properly say about this discussion?

A. Evelyn is mistaken in thinking Lou's argument to be fallacious, and so her own conclusion is unwarranted.

B. Evelyn is right about Lou's argument, but nevertheless her own conclusion is unwarranted.

C. Since Evelyn is right about Lou's argument, her own conclusion is well supported.

D. Since Evelyn is mistaken about Lou's argument, her own conclusion must be false.

E. Evelyn is right about Lou's argument, but nevertheless her own conclusion is false.

Figure 41. Reasoning question involving the fallacy of denying the antecedent.

warranted. The correct answer is B. The test administrators emphasize that while knowledge of the terminology of formal logic is not expected (you don't have to know that this is called the fallacy of denying the antecedent), a critical and full understanding of the reasoning principles is expected.

The causes of error-making in deductive reasoning are philosophical, psychological, and psycholinguistic. Erroneous deductions are due partially to language and partially to cognitive inability. Mary Henle suggested that logical forms do not describe actual thinking; rather, logical forms such as valid syllogisms represent an ideal, or "how we ought to think."[21] Perhaps we persist in faulty logic but not in faulty reasoning. In other words, people are "logical" and have an inference scheme that enables them to advance from one step to another in a deduction, but the "effective" premises used by the reasoner to arrive at an inference may not be the ones the questioner intended.[22] Whether we ignore available information or add information from our own experience, our attention is not focused on the logician's task.

In their survey on critical thinking and reasoning, Phares O'Daffer and Bruce Thornquist summarize four main reasons for errors in deductive reasoning:

- Adding to, altering, or ignoring items from the premise.
- Allowing factual content to supersede the inference pattern. Traditional patterns of everyday discourse often override logic.
- Language difficulties, number and location of negations, sentence and word length, and cognitive overload.
- Inability to accept the hypothetical.[23]

Paradoxes

Paradoxes are persons, things, or situations that exhibit an apparently contradictory nature. As such, they are not necessarily errors or fallacies but have been studied for their lack of logical consistency. Some of these paradoxes puzzled logicians long before the time of Aristotle and the Stoics.

One form of paradox belongs mainly to the class of sophisms, using language to trick the victim into making ridiculous conclusions. An early version of this paradox is called the *hooded man*: You say you know your brother; but that man who came in just now with his head covered is your brother, and you did not know him.[24] In the *hooded man*, the trick is in using the word "know" in two different ways. Another early paradox is called the *horned man*: What you have not lost you still have; but you have not lost horns; so you still have horns.[25] Here, without making the premise explicit, the trickster has forced us to accept "Either you still have a thing or you have lost it" as an instance of the law of the excluded middle. But clearly there is a middle ground; if you never had the thing, you neither lost it nor still have it.

The paradox of the *crocodile and the baby*, based on an Egyptian fable, is a sort of damned if you do and damned if you don't paradox. A crocodile snatched a baby from its mother sitting beside the Nile. The crocodile promised to return the baby if the mother will answer one question *truthfully*: "Did the crocodile intend to return the baby?" Trusting that the crocodile would be forced to keep his promise, the mother answered truthfully, "No." The crocodile then responded that he could not return the baby for if he did she would not have told the truth. (This sort of ancient sophism became known as a *crocodilite*.)

The paradox of the *heap*, or *sorites* paradox, not to be confused with the multipremise chain arguments called *sorites* or *polysyllogisms*, was a paradox concerning a heap of grain. One grain does not make a heap. If one grain does not make a heap, certainly the addition of another grain does not make a heap. Therefore two grains do not make a heap. If two do not, then three do not. . . . Therefore, 10,000 grains do not make a heap. We can arrive at the same paradox by subtracting grains. If we begin with a heap of grain and take away one grain, we still have a heap. Take away another grain, and we still have a heap. Continuing in this manner, we will be forced to conclude that one grain constitutes a heap. This is one of the paradoxes attributed to the Megarian logician Eubulides of Miletus.

In the same vein, we have the *falakros* paradox, or *the bald man*. Is a man with one hair on his head a bald man? Yes, then what about two hairs? At some point we will be forced to declare that he is not bald, but where do we draw the line?[26] The vague notions of baldness and heap are not that uncommon; most notions are vague—poor and rich, small and large, few and many, and so on. One solution to paradoxes of this sort has been to replace the two-valued logic with an infinite-valued one, like fuzzy logic. Baldness is measured in degrees and so is truth. Just as one individual can be more bald than the next, one conditional in the chain can be more true (or less true) than the next. We mistake nearly true statements for completely true ones. We may regard that "a man with zero hairs is bald" is *completely true*. However, the conditional "if a man with zero hairs is bald, then a man with one hair is bald" is *almost true*. "A man with zero hairs is bald" is slightly truer than "a man with one hair is bald."

A paradox from Epimenides the Cretan became known as the *liar's paradox* and was studied intensely by medieval logicians. Paul's epistle to Titus in the *New Testament* refers to the statement, but he apparently isn't aware of the statement's paradoxical nature.[27] Epimenides claimed, "All Cretans are liars." Can he be telling the truth? If he is telling the truth, then he is a liar—being a Cretan himself. If the statement is true then it is a lie, and if it is a lie then it appears to be true. A modern version of this dilemma, "This sentence is false," is paradoxical because if it is a true sentence then it must be false as it claims and if it is a false sentence then it must be true. The difficulty in these paradoxes resides in the fact that the claims reference themselves. "This sentence is false" is self-referencing, and "All Cretans are liars" is a claim about the statement itself when it is made by a Cretan.

The problem of self-reference appears in a paradox devised in 1908 by Kurt Grelling, a German mathematician and philosopher persecuted by the Nazis. Grelling's paradox resides in a class of self-descriptive adjectives. The adjective "short" is short and "English" is English and "polysyllabic" is polysyllabic, but "long" is not long and "German" is not German and "monosyllabic" is not monosyllabic. Grelling called adjectives that were self-descriptive *autological* adjectives while those that were not descriptive of themselves were labeled *heterological*. The paradox arises in how to answer the question: Is the adjective "heterological" heterological? If it is heterological then it describes itself and so it is autological. But if it is autological, then it does describe itself, therefore "heterological" is heterological. W. V. Quine described this class of paradoxes as *antinomies* and noted that they bring on the crises in thought.[28]

Bertrand Russell, one of the most important logicians of the twentieth century, is responsible for several paradoxes that have perplexed us. Let's examine the paradox of the *village barber*: In a certain village, there is a man who is a barber; this barber shaves all and only those men who do not shave themselves. Does the barber shave himself?[29] Of course, he cannot shave himself since he shaves only men who do not shave themselves. And if he doesn't shave himself then he is one of the men that he ought to be shaving. W. V. Quine pointed out that the absurd conclusion "that he shaves himself if and only if he does not" is created by accepting the premises in the first place. No such village with such a barber can exist.

Russell introduced another paradox, named for a librarian to whom he attributed it, *Berry's paradox*. The subjects of the paradox are syllables and numbers. 2 has a one-syllable name, "two." 77 has a five-syllable name, "seventy-seven." 1,495,832 has a 17-syllable name, "one million four hundred ninety-five thousand eight hundred thirty-two." We could continue and surely find a group of numbers that defy description in fewer than 19 syllables. Berry's paradox asks us to consider the smallest such number, that is, "the least number not specifiable in less than nineteen syllables." Count the number of syllables inside the quotation marks; we have just specified that number in 18 syllables.

Russell's most famous paradox, which bears his name, was invented in 1901 and belongs to the class of antinomies. It has to do with self-membership of classes or sets. Some sets contain themselves as members; some do not. The set of all color names is not a member of itself; it is not a color name, it is a set. On the other hand, "self-swallowing" sets are members of themselves. The set of all sets with more than four members has itself more than four members and will therefore be a member of itself. Let us consider the set of all sets that are not self-swallowing, that is, the set of all sets that are not members of themselves. Does *that* set contain itself? If it is not a member of itself, by its very definition it qualifies to be a member and should be. It qualifies to be a member of itself if and only if it is not a member.[30] Russell's paradox was the first of a myriad of paradoxes that were eventually uncovered in the mathematics of set theory.[31]

Charles Sanders Peirce argued that the very first lesson that we have a right to demand from logic is how to make our ideas clear. "To know what we think, to be masters of our own meaning, will make a solid foundation for great and weighty thought."[32] Logic, reason, meaning, and thought all depend on some form of communication, even if we are communicating only with ourselves, and communication depends on language. Next we examine the problems created by language.

12
COMMON LOGIC AND LANGUAGE

Logic has a tendency to correct, first, inaccuracy of thought, secondly, inaccuracy of expression. Many persons who think logically express themselves illogically, and in so doing produce the same effect upon their hearers or readers as if they had thought wrongly.

AUGUSTUS DE MORGAN

The science of logic set out to provide us with a sound theory of reasoning, but much of the time our ability to reason logically is hampered by language. Some have argued that natural language and the language of logic follow two entirely different sets of rules. The problems that language brings to our capacity to be logical are associated with meaning, context, our cultural knowledge, and our ability to communicate through writing or conversation.

Traditionally, the art of communication fell into three categories—logic, rhetoric, and poetry—with the lines between rhetoric and logic oftentimes blurred. During the Renaissance, logic, or *dialectic*, was concerned with statements aimed at achieving valid inferences about reality and was widely considered to be the domain of scholarly and scientific discourse.[1] "Rhetoric" (far from the connotation often given the word today equating "rhetoric" with "empty rhetoric") was regarded as the method of discourse intended to communicate ideas between the learned and the populace. From the time of Zeno through the Middle Ages, logic was symbolized by the closed fist representing the tight discourse of the philosopher, while rhetoric was symbolized by the open hand representative of the open discourse between the cultured orator and the populace. Renaissance author and the "father of deductive reasoning," Sir Francis Bacon explained the difference between logic and rhetoric:

> It appeareth also that Logic differeth from Rhetoric, not only as the fist from the palm, the one close the other at large; but more is this, that Logic handleth reason exact and in truth, and Rhetoric handleth it as it is planted in popular opinions and manners.[2]

Throughout the Middle Ages, the studies of logic and rhetoric were largely confined to the universities. In 1588, shortly after the first English-language book on logic was published, Abraham Fraunce attempted to bring English law and logic together with the publication of a book on legal logic. *The Lawiers Logike* by Fraunce began with the following stanza:

> I say no more then what I saw, I saw that which I sought,
> I sought for Logike in our Law, and found it as I thought.[3]

Fraunce's work was an attempt to bring to bear the scholastic, aristocratic image of philosophy on what was considered to be at the time, the coarser, more bourgeois reputation of the law. In 1620, in an effort to do for preachers what Fraunce had done for lawyers, Thomas Granger published *Syntagma Logicum, or The Divine Logike*. Granger begins with a stanza of his own:

> This book's a *Garden* where doth grow a Tree,
> Cal'd Logike, fruitful for Theologie.[4]

One of the most popular books in logic during the seventeenth and eighteenth centuries illustrated the applications of logic to theological, civil, and ordinary discourse. *La Logique, ou L'Art de Penser* (*Logic, or the Art of Thinking*) was first published in 1662 and written by Antoine Arnauld and Pierre Nicole; both were part of a group of mystics and religious reformers that included Blaise Pascal. The book became known as *The Port Royal Logic*, named after the area near Paris where the group congregated.[5] With a wide variety of examples of common logical arguments and fallacies found in rhetoric, ethics, physics, metaphysics, and geometry, *The Port Royal Logic* was enormously popular in the late seventeenth century and remained so for two hundred years. This ambitious work discussed the operations of the mind in the formation of ideas and the meanings attached to words (semantics), while investigating the mental operations so critical to the judgment of a valid argument. Arnauld and Nicole argue that the common logic found in the world and in books of science is very different from the logic taught in schools. They underscore that common logic is not organized and arranged into neat syllogisms of universals such as those studied by the student of logic.

To a large extent, modern logic overlaps the study of *semantics*, the branch of modern linguistics that deals with the *meaning* of words, phrases, and sentences. Yet, the meanings of simple words like *and* and *or* should be clear enough, or are they? One of the easiest logical connectives to reason with is *and*; nevertheless there are a great many ways to say *and*. English words like *but*, *although*, and occasionally *while* serve to perform the same conjunctive task as *and* when the connectives contrast. "I would certainly pay you the money I owe you, *but* I haven't gotten paid yet." "*Although* I would love to go with you, I can't." "*While* you've behaved very well today, you haven't cleaned up your room." These conjunctions perform the same function as *and*, even as they alert the listener to a change of expectation.

The ambiguity of language may be the source of many deductive errors. We have seen that the meaning of the connective *or* is vague. We use it in both the exclusive sense and the inclusive sense. It has been argued that in natural language we generally use the exclusive *or*, since both of a pair of alternatives are rarely true.[6] The word *some* can take on different meanings: "at least one or possibly all" and at other times, "some but not all." Our ability to follow a line of reasoning even depends on what the meaning of *is* is.

"The car is in the garage." In this sentence, "is" provides a spatial and locative relation. "The concert is in the evening" is an example of a temporal relation. But Aristotle's "Every ***A*** is ***B***" indicates *class inclusion*. The class-inclusion relation differs from the spatial and temporal relations and has received enormous attention from students of logic and language.[7] Most examples of class inclusion involve a taxonomy: A poodle is a (type of) dog. A dog is a (type of) mammal. A mammal is a (type of) animal. The examples are hierarchical and asymmetric. The class of poodles doesn't equal the class of dogs. The class of dogs includes poodles. This relation of class inclusion between labels is prevalent in all languages, and semanticists call this relation *hyponymy*. Analogous to "synonym" and "antonym," a hyponym is a subname. Since the objects to which the word "daisy" refers are included among the objects to which the word "flower" refers, "daisy" is a hyponym of "flower."[8]

Of course, in the English language "is" can also denote identity or provide a description. Instead of saying, "London equals the capital of England," we say "London is the capital of England." We use "is" and "are" with synonyms, when the classes mutually entail each other as in the following: A duty is a tax.

A liability is a burden. Triangles are three-sided polygons. Research has shown that in syllogistic reasoning "is" is often interpreted as "is equivalent to" or "is identical with" rather than "is included in." It has been suggested that the passive voice may make the relationship clearer. Since "All ***A*** are ***B***" may suggest that ***B*** is a smaller class than ***A***, perhaps the passive voice, "All ***A*** are included in ***B***," is clearer.[9] The best of all solutions may be to return to the ordering originally used by Aristotle in the active voice: "***B*** includes all ***A***." The clarification of "is" and "are" has been shown to significantly improve performance in reasoning with logical syllogisms. Restating "All ***A*** are ***B***" as "All ***A*** are ***B*** but some ***B*** might not be ***A***," reduces the error of conversion by explicitly warning the subjects against making that all too common mistake.[10]

Within our usage of everyday language, the meanings of words depend on a whole host of factors. We automatically take *context* into consideration to understand each other and only become aware of context when something goes awry and our communication breaks down.[11] Interpreting the meaning of a statement from its context, rather than letting the words stand alone, requires an assumption about the speaker's intent. Being able to reflect on context and what we believe the speaker meant to say is precisely what makes us humans as opposed to data processors. I am constantly frustrated that my computer never knows what I meant to say or do.

Several researchers contend that there seem to be two different conventions involved in reasoning—one is that of natural language and the other that of logic.[12] We often reason based on the ground rules of natural language and practical considerations rather than reasoning via the laws of logic. The way we reason in everyday conversation is sometimes called *natural logic* or *pragmatic reasoning*.

The principal feature of our reasoning process may be its pragmatic, rather than its logical, structure. We supply premises, using known facts. We pay attention to context and we interpret meaning. In 1967, in an effort to establish the *logic of everyday conversations* H. Paul Grice, the British philosopher and logician, formulated a set of maxims that participants in any conversation implicitly follow: Be as informative as is required; do not give more information than is required; be truthful; do not say what you believe to be false or that for which you lack adequate information; be relevant; be clear; avoid obscurity of expression and ambiguity; be brief; and be orderly.[13] The *Gricean maxims* for the structure of a conversation can be summed up by one general principle: Be cooperative.

Linguists note that a conversation is not simply an exchange of information. Two people engaged in a conversation have a shared understanding of the conventions used during the process of conversing. The cooperation principle requires that the speaker try to be as informative, truthful, relevant, concise, clear, and orderly as possible, and the listener interprets what the speaker says under the assumption that the speaker is trying to be informative, truthful, relevant, concise, clear, and orderly.

If I say that some parts of the movie were interesting, it probably seems natural to accept the invited inference that some parts were not. Although logic defines *some* as *some and possibly all*, under our conventions of conversation if all parts of the movie were interesting then I should say so, since I should be as informative as possible. The convention of the conversation requires that our statements express the most comprehensive information available to the speaker.

Antoine Arnauld and Pierre Nicole expressed the cooperation principle of discourse by saying that we confine ourselves to what is actually necessary to make our meaning understood. We don't add more. This is why, in common language, premises are often suppressed; the speaker should not offer more information than is required and whatever is suppressed ought to be information available to the listener. In fact, Arnauld and Nicole indicate that it is good manners to suppress premises in conversation—a sign that the speaker takes the listeners to be intelligent enough to supply the premise themselves.

> We have already said that an enthymeme is a syllogism perfect in the mind, but imperfect in the expression, since some one of the propositions is suppressed as too clear and too well known, and as being easily supplied by the mind of those to whom we speak. This way of reasoning is so common in conversation and in writing, that it is rare, on the contrary, to express all the propositions, since there is commonly one of them clear enough to be understood, and since the nature of the human mind is rather to prefer that something be left it to supply, than to have it thought that it needs to be taught everything.
>
> Thus, this suppression flatters the vanity of those to whom we speak, in leaving something to their intelligence, and by abbreviating conversation, render it more lively and effective.[14]

This remains a feature of common logic and ordinary conversation. We assume a great deal of background and cultural knowledge when we argue or explain. In fact, we have many unspoken assumptions about how we will be understood.

Formal logic requires the reasoner to compartmentalize information (no outside knowledge can interfere) and only the minimum commitment entailed by the premises must be accepted. Formal logical reasoning requires that only those conclusions that follow from the premises be accepted, whereas everyday reasoning utilizes all of the information at the reasoner's disposal.

Another instance of the cooperation principle provides an understanding of why the conditional is often interpreted as the biconditional. If I say, "If you shovel the snow, I'll give you ten dollars," you assume that I mean, "I'll give you ten dollars if you shovel the snow and I'll give you ten dollars only if you shovel the snow." You didn't think I meant to give you ten dollars whether or not you shoveled the snow, did you? According to our rules of conversation, you assume that I am trying to be truthful, relevant, and concise; the cooperation principle practically forces you to accept the invited (yet not necessary) inference.

Studies have shown that not only are the invited inferences of the fallacies blocked by providing alternative conditions that could bring about the consequent in the conditional, but valid inferences can also be blocked by providing additional background conditions. When provided alternative conditionals and denial of one antecedent, such as "If Lisa met her friend, then she went to a play; if Lisa met her brother, then she went to a play; Lisa did not meet her friend," subjects were able to refrain from making the invited inference in the fallacy of denying the antecedent. But when provided with what the researchers called additional conditionals, rather than alternative conditionals, subjects were not able to make the valid modus ponens conclusion. For example, faced with "If Lisa met her friend, then she went to a play; if Lisa had enough money, then she went to a play; Lisa met her friend," people could not put aside their common sense, and who would expect them to? Given the additional information that money is an issue, we are invited to presume that both conditions need to be met.[15]

Several psychologists have tried to capture the logic that people actually use when reasoning. New York University psycholinguistics professor Martin Braine called this "natural logic." He explained the common fallacies that people make as an intrusion of the habits of practical reasoning and ordinary language into a task requiring formal reasoning. He cited as an example the nearly intractable problem of the notoriously poor fit between some of the connectives used in formal logic and their nearest equivalents in natural language—connectives like *if/then*, *or*, and *some*. Braine suggested that the difference between formal logic and practical reasoning has to do with the heuristic processes and comprehension strategies that individuals use rather than the logic itself.[16] Martin Gardner agrees that logic should not be confused with "heuristic reasoning"—the informal reasoning procedures that resemble the intuitive way human minds actually work when confronted with a problem.[17] We are not logic machines.

Keith Devlin, current executive director for the Stanford University Center for the Study of Language and Information, points out that computers don't have what humans use for reasoning—*commonsense knowledge*. Even when we do not speak each other's language, people are able to negotiate an understanding in a way that machines cannot. The natural languages we all speak are simply external manifestations of our common internal mental language. "Human logical thought and our use of language almost certainly involve more than the mechanistic application of rules. But that does not mean that there are no rules."[18]

The exact rules of logic may seem unimportant to some, just as there are individuals who feel that the rules of algebra or the rules of chess are of no importance to their daily lives. However, when we begin to consider whether statements do or do not of necessity follow from certain other statements, we find ourselves tackling the foundation of metaphysics, science, mathematics, epistemology, and ethics. The rules of inference and deduction are an absolute necessity for a scientific education and today, given the prevalence of computers in our lives, these rules play a particularly important role. As Mary Henle pointed out, if people were unable to reason logically, each arriving at different conclusions from the same premises, it is difficult to see how they could understand each other, follow each other's thinking, reach common decisions, and work together.[19]

13
Thinking Well—Together

Now for some people it is better worth while to seem to be wise, than to be wise without seeming to be.

Aristotle,
On Sophistical Refutations

As Princeton professor Philip Johnson-Laird has said, the business of life depends on the ability to make deductions.[1] Not only do the sciences require logic, but clear logical reasoning is also the bedrock of law and political science. The LSAT administrators state that their analytical reasoning questions "simulate the kinds of detailed analyses of relationships that a law student must perform in solving legal problems."[2] I would add that the ability to reason well is vital for solving *any* problem. The testers emphasize that the examinees should pay particularly close attention to the words used and carefully read the language to extract its precise meaning. The questions are packed with logical conditionals, conjunctions, disjunctions, and negations. The test preparation tips suggest the use of diagrams as tools for solving these questions. Figure 42 shows an example of a question in analytical reasoning.

Since only one of the answers is consistent with the rules, the other four answers must violate the rules. In addition, the question asks us to assume that W is reduced, so add W to each of the

Directions: Each group of questions in this section is based on a set of conditions. In answering some of the questions, it may be useful to draw a rough diagram. Choose the response that most accurately and completely answers each question and blacken the corresponding space on your answer sheet.

A university library budget committee must reduce exactly five of eight areas of expenditure—G, L, M, N, P, R, S, and W—in accordance with the following conditions:

If both G and S are reduced, W is also reduced.

If N is reduced, neither R nor S is reduced.

If P is reduced, L is not reduced.

Of the three areas L, M, and R, exactly two are reduced.

If W is reduced, which one of the following could be a complete and accurate list of the four other areas of expenditure to be reduced?

A. G, M, P, S

B. L, M, N, R

C. L, M, P, S

D. M, N, P, S

E. M, P, R, S

Figure 42. Analytical question from the 1996 LSAT. (*Source: The Official LSAT Sample Prep Test*, October 1966, Form 7LSS33, downloaded from *http://www.lsat.org/*. Reprinted by permission of Law School Admission Council, Inc., the copyright owner.)

five possible answers. Then the first condition, "If both G and S are reduced, W is also reduced" means we can't have G and S on the list without W. This eliminates none of the answers. The only answer with G and S is answer A, which also has W. The second condition tells us that N with R, N with S, and N with both R and S violate the condition. This eliminates answers B and D. The third condition provides the violation P with L, eliminating answer C. The fourth condition stipulates that exactly two of L, M, and R must be included and not all three. That eliminates answer A and leaves only one answer; let's check E for its consistency with the last condition. Since E does not violate that condition or any of the others, E is the correct answer.

For LSAT questions based on logical reasoning, examinees must choose the best answer from among several that could conceivably resolve the question, just as we must balance options and alternatives in our daily lives. The instructions warn us not to introduce assumptions that are by commonsense standards implausible, superfluous, or incompatible with the passage. Figure 43 displays such a question. The text is very dense but we can recognize several logical terms and concepts that have been discussed. "A lack of . . ." is a negation, as is "No scientific idea . . . ," and answers D and E utilize the quantifiers *all* and *some*. Furthermore, once again it is important that we distinguish between truth and validity. We have to assume the passage is TRUE, even if we do not agree with it. Answers B, C, and D seem to be the central thesis of the passage, so we have to agree that they are true. The opening sentence reads, "The crux of creativity resides in the ability to manufacture variations on a theme." Let's paraphrase it in the form of a conditional, "If a person has creativity then the person has the ability to manufacture variations on a theme." Answer A is the contrapositive of that conditional, "If a person does not have the ability to manufacture a variation on a theme then that person cannot be considered creative." Whether you agree with it or not, if the passage is true then its contrapositive, A, is true. The only statement not supported by the passage is E, that *some* discoveries are not variations on a previous theme.

Directions: The questions in this section are based on the reasoning contained in brief statements or passages. For some questions, more than one of the choices could conceivably answer the question. However, you are to choose the best answer; that is, the response that most accurately and completely answers the question. You should not make assumptions that are by commonsense standards implausible, superfluous, or incompatible with the passage. After you have chosen the best answer, blacken the corresponding space on your answer sheet.

The crux of creativity resides in the ability to manufacture variations on a theme. If we look at the history of science, for instance, we see that every idea is built upon a thousand related ideas. Careful analysis leads us to understand that what we choose to call a new theme or a new discovery is itself always and without exception some sort of variation, on a deep level, of previous themes.

If all of the statements in the passage are true, each of the following must also be true EXCEPT:

A. A lack of ability to manufacture a variation on a previous theme connotes a lack of creativity.
B. No scientific idea is entirely independent of all other ideas.
C. Careful analysis of a specific variation can reveal previous themes of which it is a variation.
D. All great scientific discoverers have been able to manufacture a variation on a theme.
E. Some new scientific discoveries do not represent, on a deep level, a variation on previous themes.

Figure 43. Logical question from 1996 LSAT. (*Source*: *The Official LSAT Sample Prep Test*, October 1996, Form 7LSS33, downloaded from *http://www.lsat.org/*. Reprinted by permission of Law School Admission Council, Inc., the copyright owner.)

Several notable examples of critical moments in which logical reasoning could have averted disaster have recently been brought to light. The nuclear power plant operators conducting the experiment that led to the Chernobyl disaster were faced with information not unlike: "If the test is to continue, the turbines must be rotating fast enough to generate power. The turbines are not rotating fast enough." Had the operators made the modus tollens inference, "Therefore the test should not continue," perhaps the meltdown could have been averted.[3]

Is it a question of which brain hemisphere we bring to bear when reasoning? The split-brain research from the sixties, which expounded the theory that thinking in logical categories was a strictly left hemisphere function while mental imagery and spatial awareness were handled on the right, has been largely abandoned in favor of a theory that the distinction between the two hemispheres is a subtle one—one of processing style. The left-brain areas are good at processing the precise representation of words and word sequences while the right brain supplies context and meaning. Today scientists believe that every mental faculty is shared by both parts of the brain contributing. Certainly in syllogistic reasoning, the left hemisphere kicks in, bringing to bear the logical, verbal portion of our brain, often aided by spatial imagery and mental diagrams invoked by the right cerebral hemisphere.[4]

For the past forty years, we have assumed that people everywhere possess universal modes of thinking, such as categorization and logical reasoning. Indeed, the widespread exposure to modern Western-style education assures the introduction to logical principles. As early as the tenth century A.D., Arabic scholars were great students of Aristotelian logic and transmitted their knowledge of Aristotle to the Moslem world.[5] Buddhist logic developed its own system of syllogisms in the sixth and seventh centuries, syllogisms that appear remarkably Aristotlian, and the Buddhist logic even developed its own versions of modus ponens and modus tollens.

Because of the pervasiveness of Western-style logic, research into the universality of cognitive processes has been generally neglected. Recent research by Richard Nisbett of the University of Michigan in Ann Arbor suggests that Eastern and Western frameworks for reasoning may differ substantially. Nisbett and his colleagues indicate that Western logic, evolving as it did from the Greek tradition of open debate in the *forum*, is a more rule-based perspective, whereas Eastern culture informed from the society of Confucianism encourages social harmony rather than open debate. This cultural orientation was revealed in qualitative differences in a number of cognitive process experiments. However, when cultural orientation was not an issue—and abstract material was used—Korean and American students performed equally well on logical tasks. Nisbett notes that his work does not conclusively show that culture rules reasoning styles, but it does suggest that universal features of thought may be hard to pin down.[6]

The experts disagree and the research on the universality of reasoning processes among cultures is scant, but it appears that we must have a common underlying set of sense-making rules. If we want to make sense to each other and to ourselves, we must at least have an understanding of what it means to be consistent and what it means to be contradictory. But what about rules of deduction? Do we reason with syllogisms?

Doesn't the most remote primitive farmer make inferences about his crops and cattle? As he sends his cows out to pasture, he must reason, "If they are not hurt or sick, every cow will come home tonight." Suppose the cows are not hurt or sick. Modus ponens conclusion: "Then they all come home." Suppose a cow does not come home. Modus tollens conclusion: "Then the cow must be hurt or sick." Suppose they all come home. The farmer does not fall into the fallacy of affirming the consequent; he will examine the cows thoroughly—they may be hurt or sick nonetheless. I think we must conclude that we do indeed use the rules of inference set down so long ago.

Undoubtedly, the farmer has no explicit awareness of his inference rules. In fact, most of us have no awareness of the rules we use for reasoning. But that doesn't mean there are no rules. Gerd Gigerenzer, director of the Center for Adaptive Behavior and Cognition at the Max Planck Institute in Munich, says that intelligent decisions get shaped by the desire to maintain consistency, to revise thinking in the face of new information, to reach a swift verdict, or to make a judgment that can be justified afterward. Optimal judgments take into account all pertinent and available information. However, individuals do not usually possess the time, knowledge, or computational ability to reason "optimally." We may often pick the first satisfactory option out of many choices instead of waiting to survey all possible alternatives.[7]

Gigerenzer seeks to determine the simple psychological principles that minds actually use. He is convinced that the principles are rational in that they can be accurate and work quickly. He maintains that the mind did not evolve to perform calculations like symbolic logic and complex probability computations, but instead relies on simple thinking mechanisms that operate on available information from its surroundings.

Are the laws of logic the laws of thought? George Boole entitled his historic work *Laws of Thought*. *The Port Royal Logic* was originally entitled *The Art of Thinking*, and in the second and later editions, the authors note that some have objected to the title:

> We have found some persons who are dissatisfied with the title, *The art of thinking*, instead of which they would have us put, *The art of reasoning well*. But we request these objectors to consider that, since the end of logic is to give rules for all the operations of the mind, and thus as well for simple ideas as for judgment and reasonings, there was scarcely any other word which included all these operations: and the word *thought* certainly comprehends them all; for simple ideas are thoughts, judgments are thoughts, and reasonings are thoughts. It is true that we might have said, *The art of thinking well*; but this addition was not necessary, since it was already sufficiently indicated by the word *art*, which signifies, of itself, a method of doing something well, as Aristotle himself remarks. Hence it is that it is enough to say, the art of painting, the art of reckoning, because it is supposed that there is no need of art in order to paint ill, or reckon wrongly.[8]

Logic may, however, be the *idealization* of thinking well in natural language—serving as a clear indicator of how people should think. If logic is thought of as a model of reasoning, much like a mathematical model, logic would shed light on our thinking without being identical to it in all respects. Aristotle thought that logic was the description or model of how we reason when we reason *well*.

That is not to say that we are or should be reasoning machines. Human minds are capable of considering a great deal more than a computer can. If I am truthful and I say that the farm is in the room, most persons are capable of visualizing a perfectly logical scenario that can fit those facts—like a child's play farm in the bedroom. When proceeding on logic and definitions alone, a computer might have difficulty.

Do we have a common underlying mental language? If there are rules of human reasoning, what are they? Are they the rules of inference that we have been discussing, or some other set of rules? When you say to yourself, "That doesn't make any sense," what do you mean?

Theories of Reasoning

The idea that human reasoning at its highest level depends on formal rules of inference goes back to the ancient Greeks and was advanced by the learning theories of Jean Piaget. Piaget described the natural evolution of the reasoning process from childhood to adulthood as passing through a series of stages, the last of which we attain in our late teens. At this stage we have reached our formal reasoning period and should be able to reason about abstract material. Today, two main theories try to explain the underlying mechanisms of deduction—the theory that reasoning is based on rules and the theory that reasoning is based on models. The rule-theorists are convinced that humans reason from a collection of rules similar to the formal rules of logic although we are not necessarily aware of what those rules are. They argue that humans have a natural logic.

The mental-model theory suggests that reasoners use the meaning of premises and general knowledge to imagine the possibilities under consideration. According to the theory, reasoners build mental models based on their understanding of the premises and any relevant general information that may have been triggered during the thinking process. Reasoners will then formulate a conclusion based on their mental model and search for alternative models in which the tentative conclusion is false (searching for a counterexample). If the counterexample produces an alternative model, the step is repeated, but if the reasoner is unable to produce an alternative model the conclusion is accepted as following from the premises.

Both theories predict that the more steps that are involved in reasoning (either more models must be constructed or more rules have to be invoked to derive an inference), the more difficult a reasoning problem will be due to the restrictions on our short-term or working memory and to mental overload.[9] In many ways, rules and models are not incompatible.

A third theoretical perspective offers us an explanation as to why some inferences come "naturally," namely those that draw on the practical uses of language in our everyday lives and make sense to us. The pragmatic theorists point out, for example, that humans can reason from conditionals that give permission, define an obligation, or agree to a contract, such as "If a person is drinking beer then the person must be over 18" or "A person may drink beer only if the person is over 18."[10]

All of the theories recognize the force of a counterexample. But researchers are not exactly sure how and when we call forth counterexamples in reasoning. Unfortunately, much of the literature raises more questions than it answers. Experts agree that theories of reasoning must address the processing and interpretive assumptions in reasoning if a better account of how human beings make inferences is to be forthcoming.[11] What factors affect the ease of constructing counterexamples? When do people fail to look for counterexamples? When do they settle for a conclusion?

If folks don't usually follow the rules of logic but have some other set of natural logic rules, why don't we design a different system of logic with rules that come more naturally? But how would we explain that new set of rules to each other? We would have to use *words*—words that mean what? To express any new rule for our new logical language, wouldn't we have to use words that had the same meanings as *all*, *not*, and *if*? To explain these rules to one another what meaning would our words have? Surely they would have the logical import given to them by the current conventions of logic. So if the words and expressions we use have perfectly good meanings, why are we inventing new ones?

Even if we are reasonable, we do make reasoning errors, and we make lots of them. Antoine Arnauld and Pierre Nicole suggested that these errors arise mainly from reasoning from wrong principles rather than wrong reasoning, in other words from the *matter* being reasoned about, rather than the *form* of thought. However, Gottfried Leibniz took exception to their statements, saying that he had often observed that mathematicians themselves frequently neglected and failed in the form of thought. One translator of *The Port Royal Logic* notes that sometimes we reason logically from wrong premises and sometimes we fail to reason logically from sound premises but often we invoke unsound reasoning from unsound premises:

> There can be little doubt that ordinary reasonings fail, sometimes in the one respect, sometimes in the other, and often in both; that sound judgments often form the basis of unsound reasonings; that judgments are often unsound, while the reasonings which proceed from them are valid; and that not unfrequently the judgment is false and the reasoning vicious, at the same time.[12]

Certain subjects, such as physics, chemistry, mathematics, and medicine, may be very unintuitive to most of us. If a colleague from the chemistry department tells me that "If the substance is chloride ion, then the reaction will be a white precipitate," I would be inclined to take her word for it. I could easily, having little intuition about chloride ion and white precipitate, make appropriate modus ponens and modus tollens inferences and avoid the usual fallacies. However, when we know something about the content of a conditional, reasoning can go right out the window.

Rationality processes are highly dependent on content and context. Researchers should not be surprised that when faced with conditionals such as "If he is invited, Steven will go to the party" and "If the library stays open, then Elisa studies in the library all evening," people bring their commonsense background knowledge to the task at hand. People often tend to infer that the statements mean "Steven will go to the party only if he is invited" and "Elisa studies in the library all evening only if it is open." The antecedents are seen as necessary and with good reason. We do have some intuition about how social invitations and libraries work; we don't generally crash parties or libraries. When faced with material that conflicts with our background assumptions, what rational being would abandon common sense and strictly apply the laws of logic?

Reasoning with conditionals about what is permissible is not only sensitive to the material being used but is also sensitive to the point of view being taken. We are not likely to interpret the statement "If a person is drinking beer then the person must be over 18" to mean "A person is over 18 only if the person is drinking beer." However, the meaning of "If you clean up your room, then you may watch TV" depends on whose point of view you take. The mother is going to consider her rule violated if she discovers that her child does not clean up and watches TV anyway, and the child is going to consider his mother inconsistent with her rule if he cleans up and still doesn't get to watch TV. Most likely they both viewed the rule as a biconditional, "If you clean up your room, then you may watch TV, and you may watch TV only if you clean up your room."[13]

Some of the tasks psychologists have devised to measure reasoning ability appear to be ostensibly simple but require our short-term memory to compute far too much information. The Wason selection task has the appearance of being simple but, as one expert indicates, its very abstractness makes it complex.[14] Other tasks, such as the THOG problem, take us in by drawing our attention to the wrong place. We need a better understanding of where the subject's attention is focused and why it is focused there. Researchers say that we select what we believe to be *relevant* information, not all information. People consider the most plausible model rather than all possible models. People make judgments about what information is logically relevant and what is not, and people apply logical rules to aspects of the problem that seem relevant, but they do not actively seek out data that might prove to be relevant. This has implications for our critical thinking and decision making. As one researcher asks, "How can problems be presented to decision makers so as to maximize their attention to relevant data and minimize their concern with irrelevant data?"[15]

Apparently we make errors in logic because our attention is in the wrong place; we do not actively seek counterexamples to a tentative conclusion; and when our common sense conflicts with our logical sense we stick with our instincts, which are usually wrong. Is there evidence that we can become more logical? Can logic be taught? Most of the results are not encouraging. Some studies have indicated that even after an entire course in college introductory logic, students showed minimal improvement in their ability to use the conditional and biconditional.[16] Professors bemoan the fact that students find the study of systems of formal deduction alien to them, thereby making these courses quite difficult to teach. However, other studies have shown that while neither example training nor abstract rule training was effective in improving the subjects' abilities to solve particular concrete problems involving the conditional, when both forms of training were used in tandem, performance improved significantly.[17]

Richard Nesbitt and his colleagues suggest that the use of the pre-existing concepts of permission and obligation (both of which are forms of contract) is an effective way to teach the conditional. They found that what they call obligation-based training was effective in improving not only obligation-type problems but arbitrary problems as well. Permission and obligation conditionals are *if* ***p*** *then* ***q*** inferences wherein permission ***q*** is given for ***p***. "If you get a driver's license in New Jersey, then you must pass a written test." The consequent ***q*** is necessary permission but not sufficient permission for action ***p***. Passing a written test is necessary but perhaps not the only necessary condition (there is a minimum age requirement and some folks must pass a driving portion of the test). The importance of these teaching examples is that if they are constructed correctly and chosen wisely, common sense and intuition will prevent you from (mis)interpreting the conditional as a biconditional. "If you don't shut up, I'll scream" is a bad teaching example, whereas "If you are drinking beer legally, then you are over 18" is a good teaching example.

Piaget argued that we can reach the high level of abstraction necessary for solving problems in propositional operations, but that the cognitive development was spontaneous and could not be taught except through self-discovery. But if we are to learn through self-discovery, don't we need consistent unambiguous examples to discover from? Nisbett and his colleagues conclude we should draw on the informal inferential rule system that people use in solving recurrent everyday problems and that even brief training in inferential rules can enhance our ability to reason about everyday life events.[18]

▸ ▸

The human mind is pattern-seeking and structure-seeking. We look for cause and effect; we look for connections between events. Some hold that being logical is a learned drive; individuals are trained to be logical. If so, perhaps we can reconcile this notion with Piagetian learning theory. His levels of cognitive ability are guidelines for the time in our development when our brains are ready to grasp certain concepts. We can learn at the next level only if we are ready for it.

We have seen studies where adults evaluated the conditional with almost perfect competence when the necessity of the consequent was made explicit. Other studies indicated that when subjects were alerted to the necessary and non-necessary inferences of the conditional, fallacious inferences were greatly reduced. This language-intensive approach might be a particularly important one in the teaching of children whose language ability, as well as their ability to reason, is still developing.

It is very difficult to overcome the casual attitudes toward language and the linguistic habits that we have developed over a lifetime. We get away with being imprecise in normal discourse and make assumptions about how we will be understood, but we should be on our guard. The conventions and rules of everyday language, which depend enormously on context, are occasionally at odds with the language of logic. Reconciling our linguistic habits with the laws of logic can cause much confusion. However, precision of thought and clear deductive reasoning are goals we should all aspire to. As Charles Sanders Peirce stated, "For an individual, . . . there can be no question that a few clear ideas are worth more than many confused ones."[19]

It is also very difficult to criticize our own reasoning. Philip Johnson-Laird points out, "We are all better critics of other people's inferences than our own." He adds that while we recognize the force of counterexamples, we are more apt to construct models that reflect our own views than to search for counter-examples that refute them.[20] Very little is understood about how people discover counterexamples or recover them from memory. But exposure to searching for counterexamples is an essential ingredient in becoming more logical. We routinely make inferences from information retrieved from long-term memory, beliefs, new knowledge, and guesses. Somehow these inferences coexist with our ability to make logical and analytical decisions. We can be rational if we really put our minds to it, but ordinarily we employ lots of shortcuts that save us time and effort.

As informed citizens and intelligent human beings, we must be able to recognize the point or issue of an argument or dispute. We must be able to draw reasonable conclusions from given evidence and identify conflicting facts or arguments whether we find them in the humanities, social sciences, natural sciences, politics, religion, or the law. We must be able to understand, analyze, and criticize speeches, advertisements, newspaper articles and editorials, oral arguments from the talking heads on TV, informal discussions and conversations, as well as assertions and positions that might be put forward by experts and teachers.

We don't normally have the time to carefully analyze all of the elements of every argument presented to us, so we must decide which parts of the argument to pay close attention to. But we can learn to pay attention to the right things. When we instruct others or explain ourselves, we can be clear and use precise language. As we learn to be more logical and make our ideas clear, we are on the road to thinking well.

Notes

Introduction

1. Wason and Johnson-Laird (1972), p. 50.
2. In disjunctive concept attainment tasks, the common error made in this task is known as the common-element fallacy, according to Wason and Johnson-Laird (1972), p. 51.
3. Educational Testing Service (1992).
4. Twenty-three out of twenty-five questions had 32 percent or greater correct responses (ETS 1992). One other question had a 16 percent correct response, but in my opinion it was a trick question.
5. *http://www.gre.org/cbttest.html#description*.
6. Euler (1997), CV, p. 469.
7. Hofstadter (1979), p. 19.
8. Rucker (1987).
9. Howell (1961), p. 49.
10. Howell (1961), p. 14.
11. Bell (1937), p. 433.
12. Rucker (1987), p. 209.
13. Carroll (1896), p. 53.
14. Henle (1962); Baron (1988); Anderson (1990).

15. Staudenmayer (1975); Johnson-Laird and Wason (1977b); Griggs and Cox (1982).

CHAPTER 1

1. Aristotle (1942 trans.), *Topica* vol. I.1, p. 100a18.
2. Plato (1965).
3. Aristotle (1942 trans.), *Anal. Pr.* vol. II.2, p. 53b12.
4. Dunham (1997), p. 7.
5. Kneale and Kneale (1962), p. 8. They indicate that *reductio ad impossibile* may have been suggested to Zeno by Pythagorean mathematics.
6. Kneale and Kneale (1962), p. 3; Bell (1937), p. 20.
7. *On Interpretation* XII (22a10) translations by Cook (1938) and Ross (1942). "Contingent" means uncertain, fortuitous, or unpredictable. "Necessary" means inevitable or certain.
8. Mill (1874), p. 526.
9. Politzer's actual task has been modified but without changing the essential components.

CHAPTER 2

1. Introduction to Aristotle's *On Interpretation*, translated by Harold P. Cook (1938b), p. 7.
2. Other systems of logic may allow propositions to have more than the two values of true or false.
3. O'Brien et al. (1989).
4. Of course, the singular class can be empty if Socrates never existed or if I do not exist.
5. Inhelder and Piaget (1964), p. 64.
6. Bell (1937), p. 152.
7. Euler (1997), Introduction by Hunter.
8. Martin Gardner (1982) calls the Venn diagram an ingenious improvement.
9. Euler (1997), CII, pp. 452–53.
10. Euler (1997), CIII, p. 454.
11. Jansson (1974).
12. Epp (1999), p. 191.

13. Why should we prefer the statement "All my Ferraris are red" to be true rather than false if I have no Ferraris? If we declare it to be true, then its contradictory should be false. The contradictory of "All my Ferraris are red" is "Some of my Ferraris are not red," which declares that there are *some*. Since there are not any (red or otherwise), it is reassuring that the contradictory statement is false. If the contradictory is false, then the original statement must be true!

CHAPTER 3

1. Kneale and Kneale (1962), p. 21.
2. Wason and Johnson-Laird (1972), p. 40.
3. Wason and Johnson-Laird (1972), p. 9.
4. Wason and Johnson-Laird (1972).
5. Wason and Johnson-Laird (1972), pp. 26–27; Johnson-Laird and Byrne (2002).
6. Wason and Johnson-Laird (1972), pp. 25–26.
7. Schaeken and Schroyens (2000); Schroyens, Schaeken, Fias, and d'Ydewalle (2000).
8. Wason and Johnson-Laird (1972), pp. 30–32.
9. When meaningful material was substituted—for example, "Men outnumber women" and "Women do not outnumber men"—similar results were obtained (Wason and Johnson-Laird 1970, pp. 34–36).
10. Wason and Johnson-Laird (1972), p. 164; Braine (1978).
11. *The New York Times Magazine*, March 10, 2002, p. 20.
12. Kneale and Kneale (1962), p. 147.
13. Schaeken and Schroyens (2000); Schroyens, Schaeken, Fias, and d'Ydewalle (2000).

Chapter 4

1. Rucker (1987) and Revlis (1975).
2. This is the illustration used by Ross, translator of *The Students' Oxford Aristotle*.
3. Euler also used an asterisk in graphing several propositions together.
4. Revlis (1975), p. 107.
5. Baron (1988), p. 137.

6. Wason and Johnson-Laird (1972), p. 163.
7. Johnson-Laird (1975), p. 37; Wason and Johnson-Laird (1972), p. 164. They report that Clark's hypothesis was that the smaller the scope of the negative, the easier the statement will be to understand.
8. Earl Stanhope actually preceded Hamilton in quantification of the predicate, but his work remained unpublished until 1879, sixty-three years after his death (Harley 1879).
9. Kneale and Kneale (1962), p. 350.
10. De Morgan (1966), p. 155.
11. Peirce (1933, vol. 3, p. 481) said that since "not none" means "some," then "some-some" should mean "any."

Chapter 5

1. Sextus Empiricus (1933, trans.), commentary by Bury, p. xv.
2. Aristotle (1938c trans.).
3. Aristotle (1938c trans.), vol. I.1, p. 24b18.
4. Euler (1997), CIV, p. 464.
5. The predicate term is also called the "attribute."
6. The subject of the conclusion is called the *minor term* and the predicate of the conclusion is the *major term*. The premise that contains the major term is called the *major premise* (the first premise in each figure) and the premise that contains the minor term (the second premise) is called the *minor premise*. But it's not really necessary to know all of this to get the idea.
7. Aristotle originally classified his syllogisms into three figures according to whether the middle term concerned all or part of its class as compared to the other terms. Later, logicians classified syllogisms according to the *position* of the middle term, adding a fourth figure (Gardner 1982, p. 33).
8. Kneale and Kneale (1962).
9. Aristotle (1938c trans.), vol. II.2, p. 53b20.
10. This is the only figure that produces conclusions of all four types. The valid moods in the second figure all have negative conclusions and the valid moods in the third figure all have particular conclusions.
11. Quine (1959), p. 74.
12. Johnson-Laird and Wason (1977a), p. 86.

13. Aristotle (1938c trans.), vol. I.2, p. 25a5.
14. Aristotle also argued that "Every good is pleasurable" could be translated into "Some things pleasurable are good," giving existential import to the "every" statement. Only later did logicians think it important to consider the possibility of the empty set.
15. Kneale and Kneale (1962), pp. 231–32; Rucker (1987), p. 205.
16. These are Aristotle's 14 valid moods of the first three figures (separated by semicolons). The first figure contains six additional mood-names that Tredennick (Aristotle translator) says are *indirect* and Kneale and Kneale (1962) say were called the *subaltern moods*. Tredennick gives the mnemonic mood-names for the fourth figure: Bramantip, Camenes, Dimaris (or Dimatis), Fesapo, and Fresison. To create a nice balance of six valid moods per figure, later five additional moods were added that require existential import (not allowing the empty set in the universal). They are 1:**AAI** (Barbari), 1:**EAO** (Celaront), 2:**AEO** (Camestros), 2:**EAO** (Cesaro), and 4:**AEO**. Assuming class membership (existential import), 24 of the 256 combinations are valid. If we take the narrower view that a class may be empty, only 15 are valid. If we take a still narrower view that any **E** or **I** propositions with subject and predicate terms interchanged *is a duplication* then there are even fewer valid syllogisms.
17. Halmos and Givant (1998), p. 131.
18. Leibniz (1903).
19. Venn (1894), p. 131.
20. This is Earl Charles Stanhope's example (Harley 1879).
21. Venn (1880).
22. Venn (1880).
23. Marquand (1881). Venn (1894), himself, suggests a tabular format for more than six terms. In 1885, Alexander Macfarlane suggested a table dubbed a *logical spectrum*. It consisted of one long rectangle (the universe), halved—to represent a term and its negation. Each half was again halved, representing the second term and its negation. This could be continued indefinitely so that a diagram for n terms would be composed of one long rectangle subdivided into 2^n smaller rectangles.
24. Carroll (1896), p. 244.
25. Wason and Johnson-Laird (1972), p. 130; Baron (1988), p. 140; Anderson (1990), p. 300.

26. Euler (1997), p. 478.
27. Revlis (1975), p. 96.
28. Example from Johnson-Laird in Falmagne (1975).
29. Revlis (1975), p. 94.
30. Wason and Johnson-Laird (1972), p. 129.
31. Howell (1961).
32. Kneale and Kneale (1962), p. 299.
33. Howell (1961), p. 62. An "inbeer" is an "accident"—meaning a non-essential property. The English has been changed to our modern spelling.
34. Kneale and Kneale (1962, p. 299) mention that three hundred years later the Dorsetshire poet, William Barnes, made a similar effort in *An Outline of Redecraft* (1880) but with no more success on the vocabulary of logicians.
35. Harley (1879), p. 200.
36. Euler (1997), CIV, p. 464.
37. Euler (1997), CV, pp. 468–69.

Chapter 6

1. Braine and O'Brien (1991); Anderson and Belnap (1975), p. 1.
2. According to Howell (1961), the story is in *Noctes Atticae*, 5.10. The story is also told by Diogenes Laertius in *De Vita et Moribus Philosophorum Libri X*, 9.56. Both tell the story of Euathlus and Protagoras. Sextus Empiricus tells the same story but with Corax, the inventor of the art of rhetoric, in place of Protagoras in *Adversus Mathematicos*, 2.96–99. Thomas Wilson, in *Rule of Reason* (1551) called Protagoras, Pythagoras.
3. Klein (1975), p. 181.
4. To see an example of Winkler's system, visit the website: *http://cm.bell-labs.com/cm/ms/who/pw/cryppie.htm*.
5. Wason (1977).
6. Durand-Guerrier (1996).
7. O'Brien (1987), p. 77; Evans (1984).
8. Wason and Johnson-Laird (1972), p. 199.
9. Durand-Guerrier (1996).
10. Osherson (1975), p. 89.
11. Henle (1962), p. 375; O'Brien (1987), p. 78.
12. Described in Johnson-Laird and Wason (1977b), p. 151.

13. Griggs and Cox (1982).
14. O'Brien (1987), pp. 81–82 in a second study by Cox and Griggs (1982).
15. Staudenmayer (1975).
16. O'Brien (1987).
17. Epp (1999).
18. Sherwood (1964).
19. Wason and Johnson-Laird (1972), p. 61.
20. Cassells, Schoenberger, and Grayboys (1978).
21. Koehler (1996); Eddy (1982).
22. Koehler (1996).
23. Wason and Johnson-Laird (1972), p. 61.
24. Politzer (1981).
25. O'Brien (1987); O'Brien et al. (1989). The experiment is far more complicated than this, but this is the essential component for our current interest.
26. Wason and Johnson-Laird (1972), p. 147.
27. O'Brien (1987), p. 76; Wason and Johnson-Laird (1972), p. 64.
28. Wason and Johnson-Laird (1972, p. 65) explain: If p is the unique cause of q, then when p does not occur the cause of q is absent and q will not occur either. When q does occur, its unique cause, p, must have occurred.
29. Wason and Johnson-Laird (1972), p. 73.
30. Many thanks to Dr. Treva Pamer, Chemistry Department, New Jersey City University.
31. Braine (1978), p. 8.
32. Wason and Johnson-Laird (1972), p. 73.
33. Wason and Johnson-Laird (1972), pp. 81 and 82.
34. Wason and Johnson-Laird (1972), p. 83.
35. Kneale and Kneale (1962), pp. 99 and 134.
36. Aristotle (1942 trans.), vol. II.2, p. 53b12.
37. Rumain, Connell, and Braine (1983).
38. Baron (1988), p. 137.

CHAPTER 7

1. Kneale and Kneale (1962), p. 128.
2. Although *or* connects *propositions*, "I will get an A in math or history" is easily translated to "I will get an A in math or I will get an A in history."

Furthermore, we are not discussing *questions* but *complex propositions*. However, we find the same interpretation in propositions. "Coffee or tea?" is short for "Do you want coffee or do you want tea?" and "I will make you coffee or tea" is short for "I will make you coffee or I will make you tea." I probably mean NOT BOTH coffee and tea. Some examples are from Hersh (1997).

3. Kneale and Kneale (1962), p. 147.
4. Wason (1977), p. 126.
5. Johnson-Laird (1975).
6. Osherson (1975), p. 85; Wason and Johnson-Laird (1972), p. 71. Recall the difficulty of the experiment in Chapter 1 involving a disjunction inside a biconditional—"*If* a symbol has either the color I am thinking about, *or* the shape I am thinking about, or both, *then* I accept it, but otherwise I reject it."
7. Wason and Johnson-Laird (1972), pp. 61–62. Galen called the inclusive "or" the *quasi-disjunctive*.
8. Kneale and Kneale (1962, p. 182) say that the term "hypothetical" first appeared in Galen's writing. Others used the word *hypothetical* to refer to the *conditional* only.
9. These schema are recognized by ancient authorities: Sextus Empiricus (1993 trans.), vol. I.2, p. 157 and vol. IV.8, p. 224.
10. Kneale and Kneale (1962), p. 159.
11. O'Brien (1987).
12. Devlin (1997) tells the story of how The Logical Theorist produced by Allen Newell, Clifford Shaw, and Herbert Simon of RAND Corporation fits into the effort to create machines that think.
13. Wason and Johnson-Laird (1972); Braine (1978); Johnson-Laird and Wason (1977a), p. 83.
14. Stcherbatsky (1962).
15. Politzer (1981), p. 462.
16. This fallacy is also called *obversion*.
17. Braine (1978); Braine and O'Brien (1991). Research has shown that the most and the least sophisticated subjects get modus tollens correct and that for individuals with no formal training in logic correct performance in modus tollens *decreases* with age.
18. Rumain, Connell, and Braine (1983); also see Dieussaert, Schaeken, Schroyens, and d'Ydewalle (2000).

19. With modus tollens, both groups performed well with the simple premise and performance declined with the expanded premise.
20. FORTRAN originally meant IBM Mathematical FORmula TRANslation System.

Chapter 8

1. Kneale and Kneale (1962), p. 185.
2. Hobbes (1930), sect. V, p. 11.
3. Peirce (1933), vol. 3, p. 402.
4. De Morgan (1966), p. 213, footnote.
5. Peirce (1933), vol. 3, p. 640.
6. The Educational Testing Service (ETS) indicates that after 25 years, in October 2002, this portion of the test is being phased out and will be replaced by the analytical writing test.
7. Miller and Johnson-Laird (1976), p. 244. They are quoting Johnson-Laird (1975).
8. Miller and Johnson-Laird (1976).
9. Borrowed from Wilcox (1979).
10. Wason and Johnson-Laird (1972), pp. 104–5.
11. Wason and Johnson-Laird (1972), p. 151.

Chapter 9

1. Leibniz (1966) mentions Thomas Hobbes's work, *Elements of Philosophy concerning Body*, in which Hobbes says that reasoning is *computation*.
2. Latta (1925), p. 85.
3. Wiener (1951), p. 16.
4. Leibniz (1966).
5. Leibniz (1966), p. 202.
6. *Latino sine Flexione*, or *Interlingua*, was the brainchild of another mathematician, Giuseppe Peano.
7. Couturat's correspondence with Bertrand Russell has come to light recently. Ironically, Couturat who was a pacifist was killed when his car was hit by the car carrying the orders for the mobilization of the French army the day war broke out.
8. He perfected his system by 1679 but published his new science of binary

numbers in *Essay d'une nouvelle science des nombres* to the Paris Academy when he was elected to its membership in 1701.

9. A picture of the Leibniz stepped drum calculating machine can be seen at these web addresses: *http://www-history.mcs.st-andrews.ac.uk/history/Bookpages/Leibniz_machine.jpeg* and *http://www-history.mcs.st-andrews.ac.uk/history/Bookpages/Leibniz_machine2.jpeg* and in color at *http://www.hfac.uh.edu/gbrown/philosophers/leibniz/Calculator/Calculator.html.*
10. De Morgan (1966), Heath's introduction, p. x.
11. De Morgan also mentions graphical representations by Johann Lambert. The inattention to Leibniz could be partly due to the rift between British and continental mathematicians over the priority of Newton's and Leibniz's calculus.
12. De Morgan (1966), p. 184 footnote.
13. Harley (1879).
14. Also called the empty set or null set.
15. Boole used the *exclusive* "or," so that in $X + Y$, X and Y were mutually exclusive or disjoint classes (no overlap in membership). Later logicians found it more advantageous to use the *inclusive* "or" and we do so today.
16. Also, De Morgan's rules were derived by Arnold Geulincx and included in his book, which was a popular text of the seventeenth century (Kneale and Kneale 1962, pp. 314–15).
17. De Morgan (1966), p. 119.
18. De Morgan (1966), p. 255.
19. Hofstadter (1979).
20. Rucker (1987).
21. The agreed-upon order rules are: the negation symbol only applies to its immediate follower and operations are performed in this order: conjunction, disjunction, conditional, and biconditional.
22. Sheffer (1913) called his operation *rejection*.
23. Gardner (1982), p. 127.

Chapter 10

1. Jevons (1870).
2. *Numerically definite* means that "some" has a number or a percentage attached to it.
3. Harley (1879).
4. Jevons (1870); Gardner (1982), pp. 91–92.

5. Kneale and Kneale (1962).
6. The abecedarium was a precursor of the truth table. Mays and Henry (1953) indicate that in 1874 Jevons considered the possibility of representing the presence and absence of a combination of terms with a 1 or 0, respectively. His views on the use of binary notation in this context are prophetic.
7. Jevons (1870); Mays and Henry (1953).
8. Venn (1880). A picture of his design can be found on page 17.
9. Gardner (1982, p. 107) says that the Marquand machine was an "ingenious arrangement of rods and levers connected by catgut strings, together with small pins and spiral springs."
10. Marquand (1885), p. 303.
11. While Boole used the exclusive *or*, Marquand used $\boldsymbol{A} + \boldsymbol{B}$ as the nonexclusive *or*. Like Boole, $\boldsymbol{AB}$ was the conjunction *both . . . and* (Marquand, 1885).
12. Peirce (1887), p. 165.
13. Peirce (1887), p. 165.
14. Some logicians (e.g., Quine 1959) prefer the method of *reduction*.
15. This example is from Marquand (1885), regarding Anna, Bertha, Cora, and Dora. However, he does not use truth tables.
16. Kneale and Kneale (1962), p. 568.
17. Braine (1978) gives, for example, E. L. Keenan's 1973 "On semantically based grammar" in *Linguist Inquiry* and G. Lakoff's 1970 "Linguistics and natural logic" in *Synthese*.
18. This is essentially the same problem we discussed earlier—that of assuming existential import when the set we speak of is empty.
19. *Stanford Encyclopedia of Philosophy* at *http://plato.stanford.edu/entries/peirce-logic/*. They cite the 1966 paper, "Peirce's Triadic Logic," by Max Fisch and Atwell Turquette in *Transactions of the Charles S. Peirce Society* 11: 71–85.
20. Belnap, N. D. (1977), pp. 8–37.
21. Peirce (1887), pp. 168–69.

CHAPTER 11

1. Mark Kantrowitz (1997) from *http://www-2.cs.cmu.edu/Groups/AI/html/faqs/ai/fuzzy/part1/faq-doc-8.html.*
2. Gardner (1982).

3. Wiener (1951), pp. 82, 87, Leibniz's *On the Logic of Probability*. In 1938, Hans Reichenbach proposed a modal logic in which true and false are replaced by probabilities from 0 to 1.
4. Bezdek (1993). I have modified the example slightly.
5. Aristotle (1955 trans.), vol. I, p. 165a15.
6. Aristotle (1955 trans.), Introduction, p. 7.
7. Aristotle classified the errors of language as ambiguity, amphiboly, combination, division of words, accent, and form of expression.
8. Aristotle identified seven such fallacies: (1) that which depends upon Accident; (2) the use of an expression absolutely or not absolutely but with some qualification of respect or place, or time, or relation; (3) that which depends upon ignorance of what "refutation" is; (4) that which depends upon the consequent; (5) that which depends upon assuming the original conclusion; (6) stating as cause what is not the cause; and (7) the making of more than one question into one.
9. Aristotle (1955 trans.), vol. V, p. 167b1.
10. Mill (1874).
11. Henle (1962).
12. Nisbett, Fong, Lehman, and Cheng (1987); Johnson-Laird and Byrne (2002).
13. Henle (1962), p. 375.
14. Wason and Johnson-Laird (1972), p. 134.
15. Braine and O'Brien (1991).
16. Wason and Johnson-Laird (1972); Braine (1978); O'Brien, Braine, and Yang (1994).
17. Wason and Johnson-Laird (1972), p. 157; O'Brien, Braine, and Yang (1994), p. 723.
18. Braine (1978); Nisbett, Peng, Choi, and Norenzayan (2001).
19. Revlis (1975).
20. Revlis (1975); Rumain, Connell, and Braine (1983); O'Brien et al. (1989).
21. Henle (1962), p. 366.
22. Braine (1978).
23. O'Daffer and Thornquist (1993), p. 46.
24. Kneale and Kneale (1962), p. 114. It is also called the *Unnoticed Man* or the *Electra*.
25. Kneale and Kneale (1962), p. 114.
26. Zalta (2002).

27. Titus 1: 12.
28. Quine (1966), p. 7.
29. Quine (1966), p. 4.
30. Quine (1966); Hofstadter (1979).
31. Among others were Georg Cantor's paradox of a set of all sets and Kurt Gödel's Incompleteness Theorem.
32. Peirce (1878).

CHAPTER 12

1. Howell (1961), p. 3.
2. Howell (1961) is quoting *The Works of Francis Bacon* (Spedding, Ellis, and Health, eds.) VI, pp. 168–9.
3. Fraunce had earlier written *The Sheapheardes Logike*, using examples from Edmund Spenser's *The Sheapheardes Calendar* (Howell 1961, p. 223).
4. Howell (1961), p. 230.
5. Arnauld and Nicole (1887 trans.).
6. Braine (1978).
7. Miller and Johnson-Laird (1976).
8. Miller and Johnson-Laird (1976).
9. Revlis (1975), p. 106.
10. Baron (1988), p. 140, cites the study by Ceraso and Provita (1971).
11. Devlin (1997).
12. Braine (1978); Politzer (1981); Baron (1988).
13. Devlin (1997).
14. Arnauld and Nicole (1887 trans.), pp. 229–30.
15. Byrne, Espino, Santamaria (1999).
16. Braine (1978).
17. Gardner (1982), p. 148.
18. Devlin (1997), pp. 167, 218.
19. Henle (1962).

CHAPTER 13

1. Johnson-Laird (1999).
2. LSAC at *http://www.lsac.org/qod/questions/analytical.htm.*
3. Byrne, Espino, and Santamaria (1999); Brooks (2000).

4. McCrone (1999).
5. Howell (1961) states that Al-Farabi's work *Liber de Divisione Scientiarum* was well known by other scholars of the period (tenth century). Other important scholars were Avicenna (eleventh century) and Averroës (twelfth century).
6. Bower (2000); Nisbett, Peng, Choi, and Norenzayan (2001).
7. Bower (1996).
8. Arnauld and Nicole (1887), pp. 14–15.
9. Mental model theories include Johnson-Laird (1975); Johnson-Laird and Byrne (2002).
10. Pragmatic reasoning theories include Griggs and Cox (1982); Nisbett, Fong, Lehman, and Cheng (1987).
11. Fillenbaum (1993); Dieussaert, Schaeken, Schroyens, and d'Ydewalle (2000); Schroyens, Schaeken, Fias, and d'Ydewalle (2000).
12. Arnauld and Nicole (1887), p. 179 and footnote to page 7 on pp. 373–74.
13. Johnson-Laird (1999).
14. Finocchario (1979).
15. Evans (1984).
16. The mean improvement was only 3 percent, as measured by the selection task.
17. Nisbett, Fong, Lehman, and Cheng (1987).
18. Nisbett, Fong, Lehman, and Cheng (1987).
19. Peirce (1878).
20. Johnson-Laird (1999).

References

Almstrum, Vicki L. 1996. Student difficulties with mathematical logic. *DIMACS Symposium Teaching Logic and Reasoning in an Illogical World,* Rutgers, The State University of New Jersey, 25–26 July. Retrieved May 2000 from *http://www.cs.cornell.edu/Info/People/gries/symposium/symp.htm.*

Anderson, Alan R., and Nuel D. Belnap. 1975. *Entailment: The Logic of Relevance and Necessity*, vol. 1. Princeton, NJ: Princeton University Press.

Anderson, John R. 1990. *Cognitive Psychology and Its Implications*. 3rd ed. New York: W. H. Freeman and Company, pp. 289–323.

Aristotle. *The Categories*. Harold P. Cook, 1938a (trans.). Cambridge, MA: Harvard University Press.

———. *On Interpretation*. Harold P. Cook, 1938b (trans.). Cambridge, MA: Harvard University Press.

———. *Prior Analytics*. Hugh Tredennick, 1938c (trans.). Cambridge, MA: Harvard University Press.

———. *The Student's Oxford Aristotle. Vol. 1, Logic. Categoriae, De Interpretatione, Analytica Priora, Analytica Posteriora.* W. D. Ross, 1942 (trans.). London: Oxford University Press.

———. *On Sophistical Refutation (De sophisticis elenchis)*. E. S. Forster, 1955 (trans.). Cambridge, MA: Harvard University Press.

———. 1994–2000. *On Sophistical Refutation (De sophisticis elenchis)*. W. A.

Pickard-Cambridge (trans.). Retrieved October 2002 from *http://classics.mit.edu/Aristotle/sophist_refut.html.*

Arnauld, Antoine, and Pierre Nicole. 1887. *The Port Royal Logic (The Art of Thinking).* 10th ed. Thomas Spenser Baynes (trans.). Edinburgh: William Blackwood and Sons. (Original work published in 1662.)

Baron, Jonathan. 1988. *Thinking and Deciding*. New York: Cambridge University Press, pp.134–67.

Barwise, Jon; Ruth Eberle; and Kathi Fisler. 1996. Teaching reasoning using heterogeneous logic. *DIMACS Symposium Teaching Logic and Reasoning in an Illogical World,* Rutgers, The State University of New Jersey, 25–26 July. Retrieved May 2000 from *http://www.cs.cornell.edu/Info/People/gries/symposium/symp.htm.*

Bell, Eric Temple. 1937. *Men of Mathematics: The Lives and Achievements of the Great Mathematicians from Zeno to Poincaré*. New York: Simon & Schuster, Inc.

Belnap, N. D. 1977. A useful four-valued logic. In J. M. Dunn and G. Epstein (eds.). *Modern Uses of Multiple-Valued Logic*. Dordrecht: Reidel.

Bezdek, James C. 1993. Fuzzy models—What are they, and why? *IEEE Transactions on Fuzzy Systems* 1 (1): 1–6.

Boole, George. 1854. *An Investigation of the Laws of Thought: On Which Are Founded the Mathematical Theories of Logic and Probabilities*. New York: Dover Publications, Inc.

Bower, Bruce. 1996. Rational mind designs. *Science News* 150 (July 13): 24–25.

———. 2000. Cultures of reason. *Science News* 157 (January 22): 56–58.

Boyer, Carl B. 1985. *A History of Mathematics*. Princeton: Princeton University Press.

Braine, Martin D. S. 1978. On the relation between the natural logic of reasoning and standard logic. *Psychological Review* 85: 1–21.

Braine, Martin D. S., and David P. O'Brien. 1991. A theory of *If*: A lexical entry, reasoning program, and pragmatic principles. *Psychological Review* 98 (2): 182–203.

Britton, Karl. 1970. *Communication: A Philosophical Study of Language*. College Park, MD: McGrath Publishing Company.

Brooks, Michael. 2000. Fooled again. *New Scientist* (December 9): 24–28.

Byrne, Ruth M. J.; Orlando Espino; and Carlos Santamaria. 1999. Counterexamples and the suppression of inferences. *Journal of Memory and Language* 40: 347–73.

Carroll, Lewis. 1889. *Sylvie and Bruno*. London: Macmillan and Co.

———. 1895. What the tortoise said to Achilles. *Mind* 4 (14): 278–80.

———. 1896. *Symbolic Logic by Lewis Carroll. Part I. Elementary*. 5th ed. *Part II. Advanced* (never previously published). Together with letters from Lewis Carroll to eminent nineteenth-century Logicians and to his "logical sister," and eight versions of the *Barber-Shop Paradox*. William Warren Bartley, III (ed.). New York: Clarkson N. Potter, Inc., 1977.

———. 1960. *The Annotated Alice. Alice's Adventures in Wonderland & Through the Looking Glass*. With an introduction and notes by Martin Gardner. New York: Clarkson N. Potter, Inc.

Carruccio, Ettore. 1964. *Mathematics and Logic in History and in Contemporary Thought*. Isabel Quigly (trans.). London: Faber and Faber.

Cassells, Ward; Arno Schoenberger; and Thomas Grayboys. 1978. Interpretation by physicians of clinical laboratory results. *New England Journal of Medicine* 299: 999–1001.

Ceraso, J., and A. Provitera. 1971. Sources of error in syllogistic reasoning. *Cognitive Psychology* 2: 400–10.

Church, Alonzo. 1956. *Introduction to Mathematical Logic. Vol. 1*. Princeton: Princeton University Press.

Cicero. *Topica*. H. M. Hubbell, 1949 (trans.). Cambridge, MA: Harvard University Press.

Clark, H. H. 1977. Linguistic processes in deductive reasoning. *Thinking: Readings in Cognitive Science*. Philip N. Johnson-Laird and Peter Cathcart Wason (eds.). Cambridge University Press, pp. 98–113.

Davis, Martin. 2000. *The Universal Computer: The Road from Leibniz to Turing*. New York: W. W. Norton & Company.

De Morgan, Augustus. 1966. *On the Syllogism and Other Logical Writings*. Peter Heath (ed.). New Haven: Yale University Press.

Devlin, Keith. 1997. *Goodbye, Descartes: The End of Logic and the Search for a New Cosmology of the Mind*. New York: John Wiley & Sons, Inc.

Dieussaert, Kristien; Walter Schaeken; Walter Schroyens; and Géry d'Ydewalle. 2000. Strategies during complex conditional inferences. *Thinking and Reasoning* 6: 125–60.

Dunham, William. 1990. *Journey through Genius*. New York: Penguin Books.

Durand-Guerrier, Viviane. 1996. Conditionals, necessity, and contingence in mathematics class. *DIMACS Symposium Teaching Logic and Reasoning in an Illogical World*, Rutgers, The State University of New Jersey, 25–26 July.

Retrieved May 2000 from *http://www.cs.cornell.edu/Info/People/gries/symposium/symp.htm.*

Eddy, David M. 1982. Probabilistic reasoning in clinical medicines: Problems and opportunities. *Judgment under Uncertainty: Heuristics and Biases.* New York: Cambridge University Press, pp. 249–67.

Educational Testing Service (ETS). 1992. *The PRAXIS Series Professional Assessments for Beginning Teachers: NTE Core Battery Tests Practice & Review.* Princeton, NJ: ETS.

———. 1997. *GRE (Graduate Record Examination) Practice General Test.* Retrieved January 2003 from *http://www.gre.org/cbttest.html#description.*

Epp, Susanna S. 1996. A cognitive approach to teaching logic. *DIMACS Symposium Teaching Logic and Reasoning in an Illogical World*, Rutgers, The State University of New Jersey, 25–26 July. Retrieved May 2000 from *http://www.cs.cornell.edu/Info/People/gries/symposium/symp.htm.*

———. 1999. The language of quantification in mathematics instruction. *Developing Mathematical Reasoning in Grades K-12, 1999 Yearbook.* Lee V. Stiff and Frances R. Curcio (eds.). Reston, VA: National Council of Teachers of Mathematics.

Euler, Leonhard. 1812. *Lettres a une Princesse D'Allemagne sur divers sujets de physique et de philosophie (de l'éloge d'Euler par Condorcet), Nouvelle édition.* Paris: Mme ve Courcier, Bachemier.

———. 1843. *Lettres a une Princesse D'Allemagne sur divers sujets de physique et de philosophie (de l'éloge d'Euler par Condorcet), Nouvelle édition.* Paris: Charpentier, Libraire-Éditeur.

———. 1997. *Letters of Euler to a German Princess on Different Subjects in Physics and Philosophy* (CI–CVIII). Henry Hunter (1795 trans.). London reprint. Bristol, England: Thoemmes Press.

Evans, Jonathan St. B. T. 1984. Heuristic and analytic processes in reasoning. *British Journal of Psychology* 75: 451–68.

Falmagne, Rachael J. (ed.). 1975. *Reasoning: Representation and Process in Children and Adults.* Hillsdale, NJ: Erlbaum.

Finocchiaro, Maurice A. 1979. The psychological explanation of reasoning: Logical and methodological problems. *Philosophy of the Social Sciences* 9 (September): 277–92.

FORTRAN information. Retrieved from *http://www-h.eng.cam.ac.uk/help/mjg17/f90/history.htm.*

Gardner, Martin. 1982. *Logic Machines and Diagrams.* 2nd ed. Chicago: University of Chicago Press.

———. 1996. *The Universe in a Handkerchief: Lewis Carroll's Mathematical Recreations, Games, Puzzles, and Word Plays*. New York: Copernicus.

Griggs, Richard A., and James R. Cox. 1982. The elusive thematic-materials effect in Wason's selection task. *British Journal of Psychology* 73: 407–20.

Gullberg, Jan. 1997. *Mathematics: From the Birth of Numbers*. New York: W. W. Norton & Company.

Gupta, S. N. 1895. Nature of inference on Hindu logic. *Mind* 4 (14): 159–75.

Halmos, Paul, and Steven Givant. 1998. *Logic as algebra*. Dolciani Mathematical Expositions, No. 21. The Mathematical Association of America.

Hammer, Eric M. 1995. *Logic and Visual Information*. Stanford, CA: Center for Study of Language and Information.

Harley, Robert. 1879. The Stanhope Demonstrator. *Mind* 4 (April): 192–210.

Henle, Mary. 1962. On the relation between logic and thinking. *Psychological Review* 69 (4): 366–78.

Hersh, Reuben. 1997. Math lingo vs. plain English: Double entrendre. *The American Mathematical Monthly* 104 (January): 48–51.

———. 1998. Math lingo. A follow-up article. Retrieved September 2002 from *http://www.maa.org/features/hersh.html*.

Hobbes, Thomas. 1930. *Hobbes Selections*. Frederick J. E. Woodbridge (ed.). New York: Charles Scribner's Sons.

Hofstadter, Douglas R. 1979. *Gödel, Escher, Bach: An Eternal Golden Braid*. New York: Vintage Books.

How Calculating Machines Work. Museum of HP (Hewlett-Packard) Calculators. Retrieved from *http://www.hpmuseum.org/mechwork.htm#stepdrum*.

Howell, Wilbur Samuel. 1961. *Logic and Rhetoric in England, 1500–1700*. New York: Russell & Russell, Inc.

Inhelder, Bärbel, and Jean Piaget. 1964. *The early growth of logic in the child*. E. A. Lunzer and D. Papert (trans.). New York: Harper & Row, Publishers.

Ishiguro, Hidé. 1972. *Leibniz's Philosophy of Logic and Language*. Ithaca, NY: Cornell University Press.

Jansson, Lars C. 1974. *The development of deductive reasoning: A review of the literature. Preliminary version*. Annual meeting of the American Educational Research Association. Chicago, IL, April 1974. (ERIC #ED 090 034).

Jevons, William Stanley. 1870. On the mechanical performance of logical inference. *Philosophical Transactions of the Royal Society* 160: 497–518.

Johnson-Laird, Philip N. 1975. Models of deduction. *Reasoning: Representa-*

tion and Process in Children and Adults. Rachael J. Falmagne (ed.). Hillsdale, NJ: Erlbaum, pp. 7–54.

———. 1999. Deductive reasoning. *Annual Review of Psychology*. Retrieved December 2002 from *http://www.findarticles.com/cf_0/m0961/1999_Annual/54442295/p1/article.jhtml.*

Johnson-Laird, Philip N., and Ruth M. J. Byrne. 2002. Conditionals: A theory of meaning, pragmatics, and inference. *Psychological Review* 109: 646–78.

Johnson-Laird, Philip N.; Paolo Legrenzi; Vittorio Girotto; and Maria S. Legrenzi. 2000. Illusions in reasoning about consistency. *Science* 288 (April 21): 531.

Johnson-Laird, Philip N., and Peter Cathcart Wason (eds.). 1977a. Introduction to deduction. *Thinking: Readings in Cognitive Science*. Cambridge: Cambridge University Press, pp. 75–88.

Johnson-Laird, Philip N., and Peter Cathcart Wason. 1977b. A theoretical analysis of insight into a reasoning task. *Thinking: Readings in Cognitive Science*. Cambridge: Cambridge University Press, pp. 143–57.

Kahneman, Daniel; Paul Slovic; and Amos Tversky (eds.). 1982. *Judgment under Uncertainty: Heuristics and Biases*. New York: Cambridge University Press.

Kant, Immanuel. 1800. *Logic: A Manual for Lectures.* Robert S. Hartman and Wolfgang Schwarz, 1974 (trans.). Indianapolis: Bobbs-Merrill Company, Inc.

Klein, Marvin L. 1975. Inferring from the conditional: An exploration of inferential judgments by students at selected grade levels. *Research in the Teaching of English* 9 (2): 162–83.

Kneale, William, and Martha Kneale. 1962. *The Development of Logic*. London: Oxford University Press.

Koehler, Jonathan J. 1996. The base rate fallacy reconsidered: Descriptive, normative, and methodological challenges. *Behavioral and Brain Sciences* 19 (1): 1–53.

Latta, Robert (trans.). 1925. Quotes from *De Scientia Universali seu Calculo Philosophico* in *Leibniz: The Monadology and Other Philosophical Writings*, Oxford University Press, p. 85.

Law School Admission Council. 1996. *The Official LSAT Sample Prep Test*, October 1996, Form 7LSS33. Retrieved December 2002 from *http://www.lsac.org.*

Leibniz, Gottfried. 1903. De Formae Logicae Comprobatione per Linearum

Ductus from *Die Philosophischen Schriften*, vol. VII, sect. B, IV, pp. 1–10. *Opuscules et fragments inédits de Leibniz* / extraits des ms. de la Bibliothèque royale de Hanovre par Louis Couturat. Paris. Bibliothèque Nationale in Paris Gallica. Retrieved August 2002 from *http://gallica.bnf.fr/scripts/ConsultationTout.exe?E=0&O=N068142*.

———. 1966. *Logical Papers*. G.H.R. Parkinson (trans. and ed.). Oxford: Clarendon Press.

Leibniz's calculating machine. n.d. Retrieved October 2002 from *http://www.hfac.uh.edu/gbrown/philosophers/leibniz/Calculator/Calculator.html*.

Leibniz's stepped drum calculating machine. n.d. Retrieved October 2002 from *http://www-history.mcs.st-andrews.ac.uk/history/Bookpages/Leibniz_machine.jpeg* and *http://www-history.mcs.st-andrews.ac.uk/history/Bookpages/Leibniz_machine2.jpeg*.

Macfarlane, Alexander. 1885. The logical spectrum. *The London, Edinburgh, and Dublin Philosophical Magazine and Journal of Science* 19: 286–90.

Marquand, Allan. 1881. A logical diagram for *n* terms. *The London, Edinburgh, and Dublin Philosophical Magazine and Journal of Science* 12 (October): 266–70.

———. 1885. A new logical machine. *Proceedings of the American Academy of Arts and Sciences* 21: 303–7.

Maurer, Stephen B. 1996. Teaching reasoning, broadly and narrowly. *DIMACS Symposium Teaching Logic and Reasoning in an Illogical World,* Rutgers, The State University of New Jersey, 25–26 July. Retrieved August 2002 from *http://www.cs.cornell.edu/Info/People/gries/symposium/symp.htm*.

Mays, W., and D. P. Henry. 1953. Jevons and logic. *Mind* 62 (October): 484–505.

McCrone, John. 1999. Left brain, right brain. *New Scientist* (July 3): 26–30.

Mill, John Stuart. 1874. *A System of Logic: Ratiocinative and Inductive. Being a Connected View of the Principles of Evidence and the Methods of Scientific Investigation*. 8th ed. London: Longman's, Green. (1941 reprint).

Miller, George A., and Philip N. Johnson-Laird. 1976. *Language and Perception*. Cambridge, MA: The Belknap Press of Harvard University Press, pp. 240–47, 492–505.

Nisbett, Richard E.; Geoffrey T. Fong; Darrin R. Lehman; and Patricia W. Cheng. 1987. Teaching reasoning. *Science* 238 (October 30): 625–31.

Nisbett, Richard E.; Kaiping Peng; Incheol Choi; and Ara Norenzayan. 2001. Culture and systems of thought: Holistic versus analytic cognition. *Psychological Review* 108: 291–310.

O'Brien, David P. 1987. The development of conditional reasoning: An iffy proposition. *Advances in Child Development and Behavior*, vol. 20. H. W. Reese (ed.). New York: Academic Press, Inc., pp. 61–90.

O'Brien, David P.; Martin D. S. Braine; Jeffrey W. Connell; Ira A. Noveck; Shalom M. Fisch; and Elizabeth Fun. 1989. Reasoning about conditional sentences: Development of understanding of cues to quantification. *Journal of Experimental Child Psychology* 48: 90–113.

O'Brien, David P.; Martin D. S. Braine; and Yingrui Yang. 1994. Propositional reasoning by mental models? Simple to refute in principle and in practice. *Psychological Review* 101 (4): 711–24.

O'Connor, J. J., and E. F. Robertson. 1996. *Couturat*. Retrieved October 2002 from *http://www-gap.dcs.st-and.ac.uk/~history/Mathematicians/Couturat.htm*.

O'Connor, J. J., and E. F. Robertson. 1998. *Leibniz*. Retrieved November 2002 from *http://www-history.mcs.st-andrews.ac.uk/history/Mathematicians/Leibniz.htm*.

O'Daffer, Phares G., and Bruce A. Thornquist. 1993. Critical thinking, mathematical reasoning, and proof. *Research Ideas for the Classroom: High School Mathematics*. Patricia S. Wilson (ed.). National Council of Teachers of Mathematics. New York: Macmillan Publishing Company.

Osherson, Daniel. 1975. Logic and models of logical thinking. *Reasoning: Representation and Process in Children and Adults.* Rachael J. Falmagne (ed.). Hillsdale, NJ: Erlbaum, pp. 81–91.

Peirce, Charles Sanders. 1878. How to make our ideas clear. *Popular Science Monthly* 12 (January): 286–302.

———. 1887. Logical machines. *American Journal of Psychology* 1 (November): 165–70.

———. 1923. *Chance, Love, Logic: Philosophical Essays*. New York: Barnes & Noble, Inc.

———. 1933. *Collected Papers. Volume 3: Exact Logic; Volume 4: The Simplest Mathematics*. Charles Hartshorne and Paul Weiss (eds.). Cambridge, MA: Harvard University Press.

Plato. 1965. *Euthydemus*. Rosamond Kent Sprague (trans.). Indianapolis: Bobbs-Merrill Company, Inc.

Politzer, Guy. 1981. Differences in interpretation of implication. *American Journal of Psychology* 94 (3): 461–77.

Post, Emil Leon. 1921. Introduction to a general theory of elementary propositions. *The American Journal of Mathematics,* xliii: 163–85.

Povarov, Gelliy N. *The First Russian Logic Machines*. Alexander Y. Nitussov (trans. and ed.). Retrieved from *http://www.taswegian.com/TwoHeaded/PartI5.htm*. (First published as Povarov, G., and A. Petrov. 1978. *Russian Logic Machines*. Moscow.)

Quine, Willard Van Orman. 1959. *Methods of Logic*. (revised ed.) New York: Holt, Rinehart and Winston.

———. 1966. *The Ways of Paradox and Other Essays*. New York: Random House.

Reichenbach, Hans. 1947. Analysis of conversational language. *Elements of Symbolic Logic*. New York: Macmillan.

Revlis, Russell. 1975. Syllogistic reasoning: Logical decisions from a complex data base. *Reasoning: Representation and Process in Children and Adults*. Rachael J. Falmagne (ed.). Hillsdale, NJ: Erlbaum, pp. 93–133.

Rucker, Rudy. 1987. *Mind Tools: The Five Levels of Mathematical Reality*. Boston: Houghton Mifflin Company.

Rumain, Barbara; Jeffrey Connell; and Martin D. S. Braine. 1983. Conversational comprehension processes are responsible for reasoning fallacies in children as well as adults: *If* is not the biconditional. *Developmental Psychology* 19: 471–81.

Schaeken, Walter, and Walter Schroyens. 2000. The effect of explicit negatives and of different contrast classes on conditional syllogisms. *British Journal of Psychology* 91: 533–50.

Schaeken, Walter; Walter Schroyens; and Kristien Dieussaert. 2001. Conditional assertions, tense, and explicit negative. *European Journal of Cognitive Psychology* 13: 433–50.

Schroyens, Walter; Walter Schaeken; and Géry d'Ydewalle. 1999. Error and bias in meta-propositional reasoning: A case of the mental model theory. *Thinking and Reasoning* 5: 29–65.

Schroyens, Walter; Walter Schaeken; Wim Fias; and Géry d'Ydewalle. 2000. Heuristic and analytic processes in propositional reasoning with negatives. *Journal of Experimental Psychology: Learning, Memory, and Cognition* 26: 1713–34.

Selden, Annie, and John Selden. 1996. The role of logic in the validation of mathematical proofs. *DIMACS Symposium Teaching Logic and Reasoning in an Illogical World*, Rutgers, The State University of New Jersey, 25–26 July. Retrieved August 2002 from *http://www.cs.cornell.edu/Info/People/gries/symposium/symp.htm*.

Sextus Empiricus. *Volume I: Outlines of Pyrrohonism; Volume II: Against the Logi-*

cians; Volume III: Against the Physicists and *Against the Ethicists; Volume IV: Against the Professors.* Rev. R. C. Bury, 1933 (trans.). Cambridge, MA: Harvard University Press (1961 reprint).

Sheffer, Henry Maurice. 1913. A set of five independent postulates for Boolean algebras, with application to logical constants. *Transactions of the American Mathematical Society* 14 (4): 481–88.®

Sherwood, John C. 1964. *Discourse of Reason: A Brief Handbook of Semantic and Logic*. New York: Harper & Row.

Smith, Karl. 1995. *The Nature of Mathematics*. 7th ed. Pacific Grove, CA: Brooks/Cole Publishing.

Staudenmayer, Herman. 1975. Understanding conditional reasoning with meaningful propositions. *Reasoning: Representation and Process in Children and Adults*. Rachael J. Falmagne (ed.). Hillsdale, NJ: Erlbaum, pp. 55–80.

Stcherbatsky, Fedor Ippolitovich. 1962. *Buddhist Logic. Volumes I and II*. New York: Dover Publications.

Tammelo, Ilmar. 1978. *Modern Logic in the Service of Law*. Wien, Austria: Springer-Verlag.

Venn, John. 1880. On the diagrammatic and mechanical representations of propositions and reasonings. *The London, Edinburgh, and Dublin Philosophical Magazine and Journal of Science* 9 (59): 1–18.

———. 1894. *Symbolic Logic*. 2nd ed. London: Macmillan and Co.

Wason, Peter Cathcart. 1977. Self-contradictions. *Thinking: Readings in Cognitive Science*. Cambridge: Cambridge University Press, pp. 114–28.

Wason, Peter Cathcart, and Philip N. Johnson-Laird. 1972. *Psychology of Reasoning: Structure and Content*. Cambridge, MA: Harvard University Press.

Whitehead, Alfred North, and Bertrand Russell. 1925. *Principia Mathematica*. 2nd ed. London: Cambridge University Press (1957 reprint).

Wiener, Philip P. (ed.). 1951. *Leibniz Selections*. New York: Charles Scribner's Sons.

Wilcox, Mary M. 1979. *Developmental Journey: A Guide to the Development of Logical and Moral Reasoning and Social Perspective*. Nashville: Abingdon.

Zalta, Edward N. (ed.). *Stanford Encyclopedia of Philosophy*. Winter 2002. Retrieved from *http://plato.stanford.edu/entries/logic-classical.* Entries: classical logic; fuzzy logic; informal logic; many-valued logic; modal logic; sorites paradox.

Acknowledgments

I would like to thank all of the families, friends, colleagues, and students who let me try logic questions on them. You know who you are. I thank New Jersey City University for providing me with the time to write this book by granting me released time from my teaching duties under a supplementary budget request and a sabbatical. I would like to thank the staff of the Congressman Frank J. Guarini Library and, in particular, reference librarian Mary Ann Rentko, who is gone but not forgotten. I would like to thank the staffs of the New York Public Library with their wonderful new library facility, SIBL. Much of my research on Leibniz's work would not have been possible without the Bibliothèque Nationale in Paris Gallica and their reproductions of Louis Couturat's publication of Leibniz's work on the Web at http://gallica.bnf.fr/. I would like to thank the folks at W. W. Norton, particularly my editor, Robert Weil, for believing in me and Brendan Curry for holding my hand. I would like to give my personal thanks to the following individuals: Ed Knappman, Winifred McNeill, Treva Pamer, and Bob Formaini.

I particularly want to thank Audrey Fisch for reading the manuscript and encouraging me. Finally, I want to thank my husband and partner, Michael Hirsch, whose tireless reading, rereading, and suggestions helped me enormously. We're in this together, honey.

Index

Page numbers in *italics* refer to illustrations.